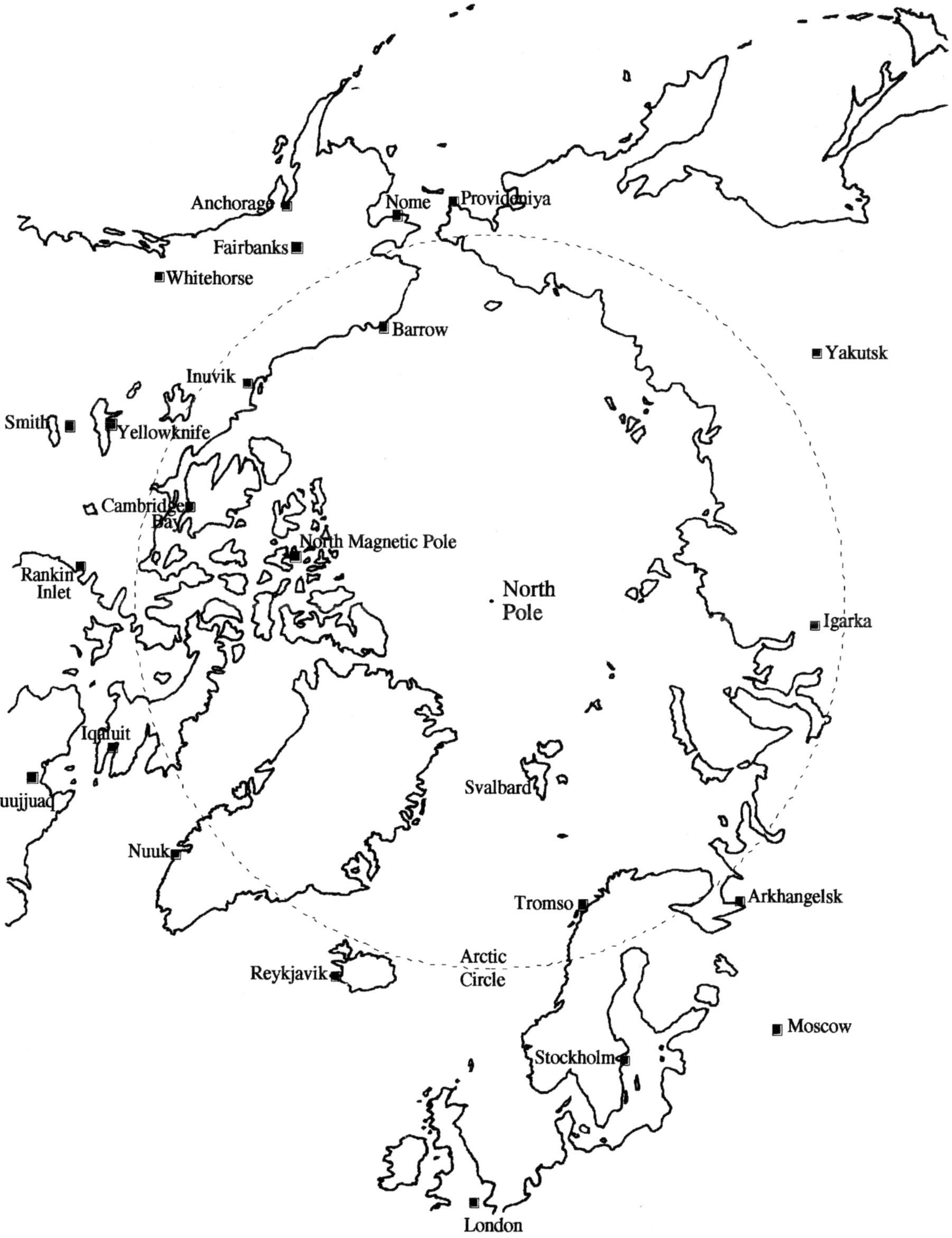
Anchorage
Nome
Providenìya
Fairbanks
Whitehorse
Barrow
Yakutsk
Inuvik
Fort Smith
Yellowknife
Cambridge Bay
North Magnetic Pole
North Pole
Rankin Inlet
Igarka
Iqaluit
Kuujjuaq
Svalbard
Nuuk
Tromso
Arkhangelsk
Arctic Circle
Reykjavik
Moscow
Stockholm
London

Some communities in northern North America

BARE POLES

BUILDING DESIGN

FOR HIGH LATITUDES

HAROLD STRUB

CARLETON UNIVERSITY PRESS

Copyright © 1996 Harold Strub
This edition Copyright © 1996
Carleton University Press

Printed and bound in Canada

Canadian Cataloguing in Publication Program

Strub, Harold, 1945-
    Bare poles : building design for high latitudes

(The Carleton library ; 185)
Includes index.
ISBN 0-88629-278-6

    1. Architecture—Arctic Regions. 2. Architecture and climate I. Title. II. Series.

NA2542.C45 1996   721'.0911   C95-920806-2

Front cover photo: Duchaussois, NWT Archives
Back cover photo: Wakenham, NWT Archives
Leaky photo, page 2: Photo by Bob Campbell.
    © National Geographic Society

Carleton University Press gratefully acknowledges the support extended to its publishing program by the Canada Council and the financial assistance of the Ontario Arts Council. The Press would also like to thank the Department of Canadian Heritage, Government of Canada, and the Government of Ontario through the Ministry of Culture, Tourism and Recreation, for their assistance.

THE CARLETON LIBRARY SERIES

A series of original works, new collections, and reprints of source material relating to Canada, issued under the supervision of the Editorial Board, Carleton Library Series, Carleton University Press Inc., Ottawa, Canada.

General Editor
    John Flood

Associate General Editor
    N.E.S. Griffiths

Editorial Board
    Pat Armstrong (Canadian Studies)
    Bruce Cox (Anthropology)
    Tom Darby (Political Science)
    Irwin Gillespie (Economics)
    Alan Hunt (Sociology)
    Peter Johansen (Journalism)
    Dominique Marshall (History)
    Iain Wallace (Geography)
    Barry Wright (Law)

The views and technology expressed in this book are not those of the Government of the Northwest Territories or its employees, and in no way reflect official government or department policies. In the design of any building at high latitudes the representations made herein must be used with care by building professionals who are aware of local context and regulation.

FOR MARI-ANN

bare poles—the condition of a ship when, in a severe storm, all her canvas has had to be taken in because of the fierceness of the wind. A ship can attempt to lie a-try under bare poles, (though she will do better under a mizzen topsail or trysail), or can scud under bare poles before the wind, very often a hazardous undertaking if there is a high sea running.

The Oxford Companion to Ships and the Sea

CONTENTS

FIGURES, TABLES AND PLATES *VIII*
FOREWORD *X*
INTRODUCTION *XI*

I PEOPLE
1. ORIGINS *3*
2. POPULATIONS *7*
3. TRANSITIONS *11*
4. LIFESTYLES *17*

II TERRAIN
5. EARTH AND WATER *23*
6. AIR AND FIRE *28*
7. FLORA AND FAUNA *33*

III CLIMATE
8. SUNLIGHT *41*
9. TEMPERATURE *45*
10. WIND, RAIN, AND SNOW *48*
11. HUMIDITY *53*

IV PROGRAM
12. SETTLEMENT *61*
13. SHELTER *66*
14. OWNER *71*
15. LOGISTICS *76*
16. BARRIERS *80*

V DESIGN
17. PLACE *88*
18. NEIGHBOURHOOD *92*
19. GEOMETRY *98*
20. PLANNING *105*
21. ENTRIES *109*
22. FOUNDATIONS *113*
23. WALLS *122*
24. WINDOWS *126*
25. DOORS *132*
26. ROOFS *135*
27. MATERIALS *141*
28. SERVICES *147*

AFTERWORD *154*
29. GLOSSARY *155*
30. BIBLIOGRAPHY *158*

APPENDICES
31. A: CLIMATE CHARTS *164*
32. B: BUILDING ELEMENTS *183*

33. INDEX *191*

LIST OF FIGURES, TABLES AND PLATES

FIGURES		PAGE
1-01	Bering land bridge	4
1-02	Linguistic groups in northern North America	9
1-06	Air circulation between the equator and the poles	28
1-08	Sunlight: angles of incidence	41
2-08	The distance sunlight travels through the atmosphere	42
3-08	Solar shading strategies vary with latitude	42
4-08	Window glass as solar heat trap	42
5-08	Daylight hours per day	44
6-08	Solar altitude at noon	44
7-08	Possible sunshine hours versus actual bright sunshine hours per month	44
1-09	Mean daily air temperature versus soil temperature	47
2-09	Degree-days below 18 degrees Celsius	47
1-10	Wind direction frequency and mean velocity at Iqaluit, September to May	51
2-10	Total precipitation	52
3-10	Mean snowfall	52
1-11	Relative humidity	56
1-13	Traditional snow house of the Canadian Central Arctic	69
1-16	Principal components of the rain screen	84
1-18	Drift pattern due to blowing snow	95
2-18	Building shaped to minimize air turbulence and drift accumulation	95
3-18	Eight rules of thumb for controlling snowdrifts around buildings (1 - 8)	96
4-18	A building form designed to control snowdrifting	97
1-19	The scale of shelter	99
2-19	Remote settlements and the scale of large public buildings	99
3-19	Perimeter-to-area ratios of building plan forms	100
4-19	Surface-to-volume ratios of building forms	100
5-19	Surface-to-volume ratios of small versus large building forms	101
6-19	Surface-to-volume ratios of single- versus two-storey detached house forms	101
7-19	Post-construction shrinkage of wood studs: screw pops	102
8-19	Post-construction shrinkage of wood studs: cracking in finished surfaces	102
9-19	Lap joints in building envelope components	102
10-19	Sealant cross section in building joints	103
11-19	Inverse square law	103
1-21	Example of a ramp up to an entrance bridge	110
2-21	Example of a wind deflector at an entrance landing	111
3-21	Example of a "floorless" wind porch	111
1-22	Frost heave as it affects steel pile foundations	115
2-22	Spreading the load	116
3-22	Foundations—seeking stability below the unstable layer	117
4-22	The thaw bulb below a heated building	117
1-23	Example of a high latitude exterior wall assembly	125
1-24	Example of a high latitude window assembly	131
1-25	Typical door supports	134
1-26	Example of a conventional roof assembly for high latitudes	139
1-28	Stack effect	149
2-28	Controlled natural ventilation	149

TABLES		PAGE
1-11	Drop in relative humidity when cold outside air is heated	55
2-11	Rise in relative humidity when room air (20% rh) is cooled	55
3-11	Rise in relative humidity when room air (40% rh) is cooled	56

PLATES	PAGE
Stone sculpture. Kimmirut	1
Footprints at Laetoli, Tanzania	2
Coming home. Gjoa Haven	2
Finger lakes. Great Bear	5
At play on shed roofs. Sanikiluaq	6
Arrival and departure. Wha Ti	8
Tomorrow's survey. Devon Icecap	10
Street scene. Cambridge Bay	13
Hard business. Repulse Bay	14
Water margin. Upernavik	15
Discussion group. Sanikiluaq	16
High-rise log cabin. Whitehorse	19
Modern cliff dwelling. Yellowknife	20
Virginia Falls. Nahanni River	21

Glaciers. Axel Heiberg Island	22
Stranded iceberg. Qeqertarsuaq	24
First snow. North Baffin Island	25
Seal on floe. Alexandra Fiord	27
Coronation Gulf. Kugluktuk	29
Horizon. Bache Peninsula	30
Aftermath of building fire. Rankin Inlet	31
Muskoxen grazing. Devon Island	34
Orchid. Nahanni River	35
Summer harbour. Ilulissat	39
Snowdrifts. Rankin Inlet	40
Disko Bay. Qeqertarsuaq	40
Roof tops. Aklavik	43
Facing the sun. Little Cornwallis Island	43
Arctic oasis. Alexandra Fiord	46
Sverdrup Glacier. Devon Island	47
Lee side snowdrifts. Rankin Inlet	49
Sastrugi. Devon Icecap	50
Truelove Lowland. Devon Island	51
100 per cent rh. Devon Icecap	54
85 per cent rh. Broughton Island	56
Hill scene. Kimmirut	59
Street scene. Nuuk	60
Zoning plan. Igloolik	60
Hudson's Bay Company. Repulse Bay	62
Street scene. Rankin Inlet	63
Key board. Chesterfield Inlet	64
Village plan overview. Povungnituk	65
Snow house. Repulse Bay	67
Thule house. Resolute	67
Neighbourhoods. Iqaluit	68
Sod houses. Uummannaq	68
Log walls for school. Lutselk'e	73
Abandoned housing. Cambridge Bay	74
Unloading the barge. Aklavik	77
Unloading the Hercules. Kugluktuk	78
Satellite dish. Upernavik	79
Exterior walls. Rankin Inlet	81
Swimming pool. Little Cornwallis Island	82
Exterior walls at Detah. Yellowknife	83
From one extreme. Gjoa Haven	86
To the other. Igloolik	86
Cluster housing. Yellowknife	87
Student residence bedroom. Rankin Inlet	87
Ancestral campsite by the water. Devon Island	88
Campsite. Kitikmeot Region	89
Village plan overview. Inukjuak	89
Modern town by the water. Uummannaq	90
Rock sways town plan. Uummannaq	92
Beach sways village plan. Hall Beach	93
Snowdrifts submerge village plan. Baker Lake	93
Snowdrift analysis. Pelly Bay	94
High-rise apartments. Nuuk	98
Wind deflector. Sanikiluaq	99
Programming model. Lutselk'e	101
Skylight at the school. Taloyoak	106
Portal frame. Ilulissat	107
Dining room. Rankin Inlet	108
Snow-free entrance. Iqaluit	109
Snow-bound entrance. Repulse Bay	110
Storage not considered. Rankin Inlet	112
New foundation for Gertie's. Dawson	113
Wood crib foundation. Taloyoak	114
Thermopile. Cambridge Bay	115
Space frame foundation. Resolute	118
Steel pipe pile foundation. Iqaluit	120
Balloon frame. Uummannaq	123
Platform frame. Rankin Inlet	124
Horizontal corrugations. Gjoa Haven	125
At home. Gjoa Haven	126
Windows and portholes. Yellowknife	127
Windows, hoods, port vents. Iqaluit	128
Skylight at the school. Taloyoak	129
Porthole. Yellowknife	130
At school. Lutselk'e	132
Going home. Rankin Inlet	133
At home. Pelly Bay	134
Roof drains onto porch. Kimmirut	135
Cold attic. Nanisivik	136
New sloped roof. Kugluktuk	137
Installing MBM roofing. Rankin Inlet	138
Thule house ruins. Skraeling Island	142
Multi-coloured wood siding. Resolute	144
Services roughed-in. Rankin Inlet	147
Kitchen from another culture. Gjoa Haven	148
Service entries stressed. Taloyoak	150
Socked smoke alarm. Rankin Inlet	151
Refuse basket. Nuuk	152

FOREWORD

Buildings designed out of context frustrate everyone—users, owners and designers. Because they don't fit the need, they are poorly cared for and do not last long. For lack of a home-grown design and construction industry, polar regions have imported copies of many such buildings from the midlatitudes over the past fifty years. Those buildings solved some problems but created others—a few passed their prime in the first five years of use. Others were derelict by age twenty. Modern buildings should have longer useful lives, and polar regions deserve better.

A home-grown design and construction industry can take generations to mature in regions as sparsely populated as the Arctic and Antarctic. The process is well underway in Canada's Northwest Territories—this book's prime example of a polar region—but a general guide for users, owners, and designers of high latitude buildings has not been published until now.

This book presents the context, in bare-bones outline, needed to design high latitude buildings with confidence. It supplies the facts and some of the questions that should be fresh in the mind of anyone making decisions about the built environment. The first three parts outline the social and physical mix of people, terrain and climate. The fourth—program—describes both potential and constraint in making buildings for people at high latitudes. The fifth part examines ways of developing potential through design at scales ranging from the small community plan to hardware for operable windows. A glossary of terms and other supporting data are located at the end of the book.

Many people have contributed comments and encouragement during the writing of this book. Gino Pin godfathered the project from start to finish. Douglas Wren jostled the author's elbow in the early stages. Barry Healey made a signal review of the first draft. Bruce Smith clarified aspects of permafrost engineering. The comments of Hilary White, Megan Williams, Dorothy Harley Eber, Martin Strube, and the late George Jacobsen were much appreciated. In 1968 the glaciologist Fritz Müller gave the author the opportunity to explore part of the High Arctic. John Flood and Jennie Strickland, sterling editors both, lent shoulders for tears and for push. Thanks also to Barbara Cumming for her typesetting and design. Thank you all.

The financial assistance of the Canada Council, and the Polar Continental Shelf Project, Government of Canada, and of the Department of Public Works and Services, Government of the Northwest Territories, is gratefully acknowledged.

Harold Strub

INTRODUCTION

I moved north to Yellowknife from Sudbury, Ontario, in October 1971. Unlike the people who have lived in the North for centuries, and earlier travellers who developed an intimate knowledge of the land by travelling its land and water routes, testing their physical and mental stamina at every step, I first saw the land from above, like most contemporary travellers. From the comfort of a Boeing 737 flying at 35,000 feet, the land below appeared to be a mosaic of earth and water textured to stimulate the senses. It was void of detail or signs of life but for the numerous seismic cut lines pushed through the Boreal forest to the horizon: the lines resembled tears in a fine tapestry inflicted by an inconsiderate invader.

In the years since then I have travelled the length and breadth of the North, becoming intimate with the land and its people. I have learned to appreciate the bursts of energy Nature uses to change the seasons: the unending winter and the not quite darkness; the rush to summer and the return of life; the first frost and the immediate jump to winter. I have marvelled at indigenous shelters refined by humans for many generations, and at the ingenious use of available materials such as snow, ice, driftwood, skins, bone, and sod to provide basic nomadic habitats. I have struggled over my own inadequate designs and agonized over our failure and that of my contemporaries to produce a comparable result. The transition from the basic nomadic settlement—a coming together of family—to the contemporary settlement—orchestrated by the planner—has not been a success.

While some attempts have been made to develop a truly polar architectural vernacular building, for instance, the new townsite at Resolute Bay by the Swedish architect Ralph Erskine, of which only a few buildings were constructed, the aborted row housing scheme for Iqaluit (Frobisher Bay) by Moshe Safdie, or the high tech reinforced fibre glass panel structures designed by Papineau Gérin-Lajoie LeBlanc, both building and community planning still follow mid-latitude design principles.

It is to this environment that Harold Strub came in the early 1980s, to join the NWT Department of Public Works, part of a government that had evolved from the planeload of bureaucrats sent from Ottawa to Yellowknife in 1967. Those were active years politically and architecturally. Self-government was moving toward reality, locally elected members were replacing a federally appointed council, federal government programs were being transferred to the new territorial government, which itself was rushing to catch up. New communities, complete with roads, power, water and sewer services, public housing, recreation and medical facilities, centred on the polarizing reality of the mission church, the Hudson Bay store, the government school, and the nursing station or the DEW Line sites, displaced the clusters of packing case shacks and federal government matchbox residences. Planned communities had arrived in the North.

This book is more than a technical chronicle of polar building. By combining technical observations and personal travel experiences, Strub has created an informative evaluation of the success and failure of recent building design in northern Canada in a unique style that is easy to relate to. It is a document that will assist both the professional designer and the inquisitive reader. It will lead, we hope, to a greater understanding of the problems of building in a polar environment and will accelerate our search for building design suited to the high latitudes.

Gino Pin, FRAIC
Yellowknife.
Former Chief Architect,
Government of the Northwest Territories.

Life-size stone sculpture. Kimmirut—April

I
PEOPLE

1 ORIGINS
NEAR THE BEGINNING 3
ICE AGES 3
ISOSTATIC REBOUND 3
THE CORRIDOR 4
PEOPLING THE AMERICAS 4
DEARTH OF FACTS 5
TIME MACHINE 5

2 POPULATIONS
ARRIVAL 7
PROFILES 7
ETHNICITY 8
LANGUAGES 8

3 TRANSITIONS
THE INVASION 11
KINSHIP 11
CONTACT 12
DEPENDENCE 13
DIRECTIONS 14

4 LIFESTYLES
SNAPSHOTS 17

*Mary Leaky measures 3.6 million
year old hominid footprints.
—Laetoli, Tanzania*

Coming home over the sea ice. Gjoa Haven—March

I ORIGINS

NEAR THE BEGINNING

In 1969 the sight of footprints on the moon made the heart skip a beat. The photographs show deep tread marks in the lunar dust next to shiny pieces of equipment. Only a generation later it is hard to fathom how the prints were made and why. In the airless sunlight they will look fresh indefinitely. In a million years they could still be marking time.

A million years? In 1976 Mary Leaky found footprints nearly four million years old. She uncovered them in an East African deposit of volcanic ash long since turned to stone. Leaky's study concluded that three human ancestors walked leisurely northward, fully upright, the setting sun on their left. An adult led the way. An adolescent trailed, taking exquisite care to place its feet inside the leader's footprints. Another adolescent skipped along to one side, pausing for a moment to look westward. Elephant, tiger, buffalo, baboon, and countless birds also left indelible footprints. Even the impact craters made by the last plump raindrops to fall before the ash solidified stare out of the ground like stopped watches.

Walking upright and playing like children that long ago suggests astonishing progress on the road to human status, a crucial step perhaps in a long series that led to the storage of surplus food, the invention of tools, the use of fire, the development of language, and the advent of questions without answers.

Still, it took almost all of those millions of years to discover the Americas. Africa we knew; Europe and Asia we got to know. But we did not trouble the New World until about fifty thousand years before present, the last percentile of our existence measured in earth years. We were rather slow on our feet.

ICE AGES

Whether 50 or 15 millennia ago, the prospect of eking out a living on the land in northeast Siberia may have discouraged eastward migration for many generations. Siberia, with Antarctica, boasts the coldest places on earth. In the northern hemisphere, on a given continent at mid- to high latitudes, a traveller moving north encounters colder winters, but so does a traveller moving from west to east. In Eurasia, Stockholm is warmer than Moscow, Moscow is warmer than Yakutsk; in North America, Fairbanks is warmer than Yellowknife, Yellowknife is warmer than Baker Lake. But once the traveller moved through the gateway, propelled from the frigid edge of the old continent to the shores of the new continent by who knows what pressures, a fresh hand was dealt: the west of the new world is warmer than the east of the old; Fairbanks is warmer than Yakutsk. The new-world may have seemed paradise to the fringe populations of Siberia.

There was another difficulty: the Pleistocene epoch, the last two and one half million years of earth history, was rife with Ice Ages, each lasting many tens of thousands of years. In areas covered by ice, human subsistence was impossible. Even the interglacial periods, thousands of years long, were far from being ice-free. A one degree Celsius drop in the continental mean air temperature was enough to trigger the growth of an ice sheet over the Laurentian Shield and reduce sea level by 150 metres. (This is Earth shifting its burden of water from one shoulder to the other, from the sea to the land.) A small rise in temperature was sufficient to force the ice into retreat northeastward (Baffin Land) and upward (the Western Cordillera) again, releasing thousands of cubic kilometres of meltwater into the sea.

ISOSTATIC REBOUND

The earth's crust floats on a sea of molten rock; the additional dead weight of an ice sheet two kilometres thick depresses land levels as much as 300 metres. When an ice sheet melts, the land surface rebounds like a car ferry being offloaded. Flush with meltwater, the sea also rises, but not nearly enough to keep up with the rise of the land freed from ice. Low-lying land never burdened by the *icecap* is inundated by the sea. As the Bering Strait opens up, two peninsulas take the place of the Siberia-Alaska isthmus. During a subsequent ice age the growing ice sheets take back water from the sea in the form of snowfall. Sea levels drop, turning the shallow shelf between the continents into an isthmus again, a land bridge for the Siberian gateway to the Americas.

Since the waves tend to swallow the evidence, no one has been able to confirm that humans crossed Bering Strait on foot whenever its surface froze solid, or crossed in skin coracles over open water. Archaeologists therefore believe that Bering Strait blocked the passage of Palaeolithic humans in migration. But the 90-kilometre gap at the strait, itself dotted with island stepping stones, seems insignificant.

To coastal peoples looking eastward it may not have been an insurmountable obstacle. To them the presence of a land bridge may have been a luxury rather than a necessity.

THE CORRIDOR

Toward the end of the last Ice Age, with the sea at low level and the land bridge in place, a migration almost certainly took place. Perhaps famine or war at home pushed people eastward. Perhaps the migration of animals reoccupying the new world terrain released by the ice sheet lured them. Once on the western edges of the new continent, the travellers encountered a milder climate. Made incrementally by successive generations the passage was not onerous. A 6,000 kilometre corridor (Figure 1-01) between the major North American icecaps (along the present-day Yukon, Porcupine, and Mackenzie valleys) connected Siberia to central North America. Incredibly, the entire evidence for this passage consists of a few crude bone tools found in river deposits of the Old Crow Flats area in north Yukon, some cutting edges made from elephant bone, a caribou bone skin scraper, and an awl that began as the leg bone of a loon.

PEOPLING THE AMERICAS

By 15,000 BP (years before present) the last Ice Age was nearing an end and, as the main ice sheets retreated, the reflooding of the Bering land bridge began. The east end of the isthmus was occupied, and a push further south started, perhaps invigorated by another wave of people from Asia. Some finds in southern California suggest that the push south may have occurred even earlier, not long after the first passage of the isthmus. In northern Chile a recent find has mummies dating from 8,000 BP, one thou-

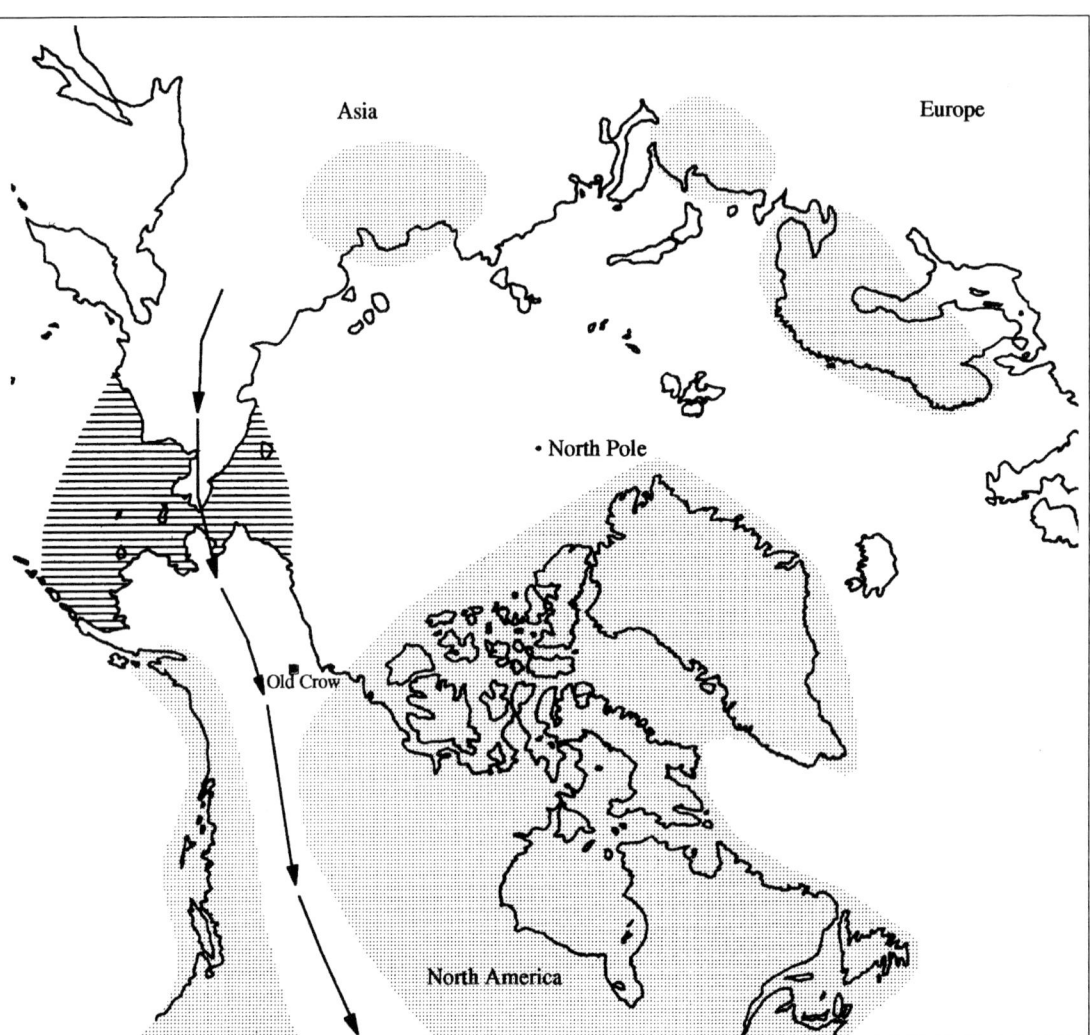

Figure 1-01 Approximate extent of the Bering land bridge and the corridor between the continental ice sheets near the end of the last Ice Age.

sand years earlier than the oldest Egyptian variety. The first reverse migration is believed to have occurred about 11,000 BP, when Amerindians from the south migrated northward into Yukon and central Alaska. Still, the lands east of the Mackenzie remained empty of humans until about 8,000 BP, when they were occupied as far north as the *tree line* by Amerindian caribou hunters from the south, properly the ancestors of the *Dene*, the "Indian" peoples of northern Canada.

The south coast of Alaska had a flourishing maritime culture by at least 7,500 BP, but the *tundra* north of the tree line in arctic Canada and Greenland remained unoccupied until about 4,500 BP, when a pre-Dorset culture designated "Independence" arrived from the west, either from Siberia or northern Alaska; these people made small tools of flaked, multi-coloured flint. By 2,500 BP this culture, having developed a richer variety of tools and artifacts, emerged as "Dorset" people (the "Tunit," according to latter-day Inuit legend). In turn the Dorset culture was submerged about 1,000 BP by another wave of migration from the west, this time a group with superior hunting techniques and an ability to hunt sea mammals large and small. These people, designated the "Thule" culture, are the direct forebears of the *Inuit*; they are thought to have had contact, perhaps indirect, with the Norse people who first made their way to North America from Iceland and Greenland about 1,000 BP. Led by Martin Frobisher, the first modern Europeans arrived at Baffin Island in 1576 AD, but did not stay; their occupation of northern parts, driven by the fur trade, did not begin in earnest until the eighteenth century.

DEARTH OF FACTS

Canadian arctic prehistory as we know it is top-heavy with theory. Facts are few because methodical excavation began only three decades ago. Of the few artifacts recovered many were found in redeposited gravels rather than in their original context, a fact that rules out stratigraphy as a determinant of age. Age

Finger lakes incised in bedrock by the continental ice sheet. Great Bear—July

determination techniques evolve quickly, recent work often undermining previous determinations. A piece of bone once thought to be 27,000 years old turns out to be 1,300 years old. Such discrepancies keep the timing of human passage between continents in turmoil. Even less work has been done in northeastern Siberia, the supposed site of New World beginnings. Perhaps the constraints of permanently frozen ground, short digging seasons, blackflies, and uncertain glory, compared, say, to the rewards of making discoveries in the Nile or Indus valleys, have kept arctic archaeology at bay. But a few more finds will double existing knowledge and lighten the overburden of conjecture, assuming that enough traces remain to form a credible picture.

TIME MACHINE

Some prehistoric evidence is plain to see. In Canada, particularly in the arctic archipelago, human and geological history converge in a remarkable way. With the end of the last Ice Age and the retreat of the icecap about 10,000 BP, *isostatic rebound* triggered reemergence of the continental land mass. Rising coastlines with shallow foreshores developed new gravel beaches every few hundred or thousand years, stranding old beaches and tidal lagoons high above sea level. In any one place along the present coast, traces of half a dozen raised beaches separated by stagnant pools, now become sedge meadow, marsh, or freshwater pond, co-exist with today's beach and breakers like so many high-water marks.

Since the beaches were still young when stranded above the waves, traces of human prehistory lie relatively close to the surface, often in plain sight. Coastal peoples here did not camp at the top of a cliff like summer cottagers, but on the beach, closest to the sea. On the highest, most inland, oldest raised beaches traces of Independence peoples are visible: a stone hearth used once or twice; a string of campsites; a tent foundation with stone hearth and rocks outlining sleeping areas and tent edge. At the next beach over, a circular depression in the beach gravel, now filled with humus, moss, and a sprinkling of flowers, probably marks the site of a pre-Dorset tent or snow house. Closer still to the sea a tight group of rectangular depressions indicates a Dorset community. Seaward some more, the beach might display a stone-lined pit with sunken entrance and raised sleeping platform: a Thule winter house, once roofed with hides stretched over whalebone rafters. And beyond this, in sight and sound of the waves, lies a not-so-old Inuit tent ring accompanied by some charred driftwood, a pair of spent .303 shells, and a rusty powdered-milk can.

Few stretches of coast display all these signs at once; human populations then as now were thin and the territory vast. Still, like a fan of cards laid face up, raised beaches tell the story of passing generations horizontally in a few instants of recognition, quite unlike soil deposits elsewhere in the world, laid down vertically, each layer a perfect mask for the time and chattels that came before. Most human beings are accustomed to walking through life oblivious to the history underfoot, but some Inuit elders can recall as children first discovering, then playing with toys and implements of civilizations long since passed into legend.

Although retired from the seaside, raised beaches are still very much alive. Plover eggs lie on them, tiny, camouflaged, invisible to the naked eye; lemming burrows and runways pierce and furrow them; owl roosts marked by flame orange lichen and bright green moss highlight them; purple saxifrage and yellow butterflies celebrate them; fox and weasel, quick on their feet, grace them; stones and rocks, weathered side up, randomly placed by the sea waves, preserve them against wind, rain, and snow.

Rocks randomly placed. Perfect contrast for rocks deliberately placed. Rocks placed by hand on this landscape stand out like torchlight in the dark. Caches, cairns, stone fox traps, *inukshuk*, tent rings, cooking fires, sleeping platforms, foundations for houses and longhouses reverberate in this part of the world until recently called silent. And they lead, unerringly, to the discovery of the arrow heads, harpoon points, awls, and like trivia that convince us of the passage and longevity of humankind. But not a single footprint remains.

At play on shed roofs. Sanikiluaq—April

2 POPULATIONS

ARRIVAL

Well below freezing the air inside the plane turns breath to fog and windows to ice. Heat from a hand pressed against the glass makes a patch of window transparent long enough for a glimpse of the snow-blown expanse of rock three hundred metres below. Evidence of the bruising contact between the continental ice sheet and the bedrock persists in massive striations running straight to the horizon, in pillowed rock fissured by freezing water, and in lakes and ponds separated by glacial till *deposited during the last ice retreat ten thousand years ago. Too smooth and exposed to nurture plant life taller than ten centimetres, the rock forces the black spruce to take root in sheltered crevices where dust, nutrients, rain, and snow drop out of the wind. Even the low stratus just above the aircraft seems to have taken its shape from the rock.*

A wedge of frozen lake comes into view, and next to it an open space covered with snow, machine tracks, and a scattering of log houses mixed with bungalows. Before joining final approach the plane banks steeply over the settlement, using engine noise to announce arrival. In the village nothing moves and chimney smoke stands upright. Finally a pickup and three skidoos strike out for the landing strip like laser beams.

The plane's skis settle onto the runway in a puff of snow, then sway and shimmy as though barely keeping up with forward momentum. A burst of reverse thrust confirms a safe landing. The aircraft turns and taxis toward the village through a blizzard stirred up by prop wash. The skidoos have already gathered where the plane comes to rest, the turbine noise dies, and a sense of relief mixed with the faint surprise of arrival flashes through the passengers as they unbuckle. By now the expectant people outside mill around the exit ready to inspect this twice-a-week distraction sent in from the "civilized" world. Adults muffled in parkas and children in windbreakers, hands in pockets, jostle for a better view.

The strangers descending the ladder into the crowd look in vain for eye contact and an assurance of welcome. But to villagers used to seeing strangers come and go, these newcomers barely rate a flicker of attention. They see kin instead and mail bags, machine parts, cases of soft drinks and potato chips. Some help unload the cargo and baggage. A few get ready to leave. Final thoughts are voiced quietly, some hands touch, and then new passengers board the plane, their brows furrowed by thoughts of tomorrow on the outside. The whine of turbines resumes and people and vehicles and plane scatter out of sight. The ten-thousand-year-old stillness, lifted so briefly, blankets the landscape again.

PROFILES

The density of that small crowd by the plane, three persons per square metre, is the equivalent of three million persons per square kilometre. But crowds are uncommon here. The density of the village is only 800 persons per square kilometre. The density of the Northwest Territories is a mere 0.019 persons per square kilometre, or about one person for every 52 square kilometres of land. Compare this to the city of Toronto at 6,300 persons per square kilometre, and to Canada as a whole at 2.5 persons per square kilometre. The population of the Northwest Territories, more than 65,000 now, spreads itself very thinly over the face of northern Canada. Caribou density is on the order of 0.25 head per square kilometre, or one caribou for every four square kilometres of land. At the moment caribou outnumber humans by 13 to 1.

A baby boom beginning at the time of permanent resettlement in the 1960s, combined with improvements in health services, means that nearly two thirds of the human population in the Northwest Territories is of working age. (Birth and mortality rates are drifting slowly downward to the Canadian national average.) Such a large work force in a land desperately short of wage employment suggests potential for social upheaval. The migration—in and out—of mostly non-native people responding to the ups and downs of the economy in other parts of Canada unsettles the unemployment situation even more. The emigration of non-natives who retire outside the Territories in places with much shorter winters contributes to the youthfulness of the resident population. Mobility—people's readiness to move elsewhere in search of better opportunity—normally rectifies inconsistencies in regional economies, but in the Northwest Territories mobility hardly exists. Native people tend to stay put near their place of birth.

Non-native populations tend to congregate in urban centres, principally Yellowknife, well connected to the mainstream economy in southern Canada by highways and non-stop air routes. Highway connections exist only in the western quarter of the Northwest Territories, the *subarctic* portion. Native

Arrival and departure. Wha Ti—April

populations tend to stay in villages close to ancestral lands isolated from mainstream Canada. The Inuit in the east and northeast have no highway connections either to the rest of the Northwest Territories or to southern Canada.

ETHNICITY

Physical geography and population distribution converge in other complex ways. The Inuit have traditionally been both arctic and coastal people, harvesters of animal life in cold sea water and on the adjacent lands north of the tree line. They live in widely separated villages along the Arctic rim, which runs east-west, not north-south. The culture of the rim makes geographic contact with other cultures on the continent only at the extremities, with the Athapaskan Dene in the west and the Algonkian Montagnais in the east. To the north the Arctic Ocean and to the south a vast expanse of uninhabited tundra isolate the Inuit from the rest of the world. Their contacts with the Dene in the west were few and transitory; some episodes ended in bloodshed. Their presence on the rim makes the present day Inuit an international people, given that national boundaries of circumpolar countries converge at the north pole. (A person standing at the north pole lives and breathes in six countries at once.) From west to east the Inuit straddle the boundaries of eastern Siberia, Alaska, Yukon, Northwest Territories, Quebec, Labrador, and Greenland.

In contrast the Dene are subarctic and riverine people, harvesters of the life found in and around rivers and lakes in the Mackenzie drainage system south and west of the tree line. The rivers provide both food and means of travel. In summer, camp members can move to new or temporary camps by canoe or skiff; logs harvested upstream can be rafted downstream to the settlement. Communication with neighbouring peoples flows north-south, parallel to the major river valleys and the Rocky Mountain chain. Populations speaking Athapaskan languages range from central Alaska through the Yukon and Northwest Territories to the northern regions of Canada's western provinces; a second "island" of Athapaskan speakers survives in the American southwest—Navajo country.

The Northwest Territories is the only Canadian territory or province to have a native majority. The Inuit form a third of the total; the Dene and the Métis—Athapaskan people whose antecedents married Euro-Canadians—form nearly a quarter. The non-native population, about forty percent of the total, mainly Canadians of European descent, are relative newcomers. Most have immigrated from other provinces since World War II, drawn in part by the rapid development of oil and mineral resources but primarily by the move of the territorial government from Ottawa to Yellowknife in 1967. The non-native population includes many Canadians with national origins as far-flung as China, Barbados, and India, who go to the Northwest Territories as entrepreneurs and professionals.

LANGUAGES

Without knowing the language an outsider cannot hope to understand the culture of another people. Despite this, few non-natives living in the Northwest Territories speak native languages. In Canada the two-way street of cultural exchange and mutual respect between peoples remains more imagined than practised. Many find the benefits of learning a second language too abstract to be worth pursuing. Aboriginal minorities view the majority's unilingualism as an obstacle to their survival as distinct peoples.

In the first half of this century the national majority imposed its language—English or French—on the native populations of Canada where the resistance was weakest and the payoff greatest: in the schooling and religious instruction of children. As a result, more than half the natives of Canada no longer speak their mother tongue. Just how does one describe the culture of persons who have never spoken their mother tongue? Is an English child brought up by grandparents in Germany English or German? What culture does an Inuk in Tuktoyaktuk represent if he or she can speak only English?

The study of the relation between thought patterns, culture, and language, still in its infancy, offers little practical assistance to the building designer trying to understand the thinking of a client living in another culture. In the simplest of examples, the word goodbye in English means "God be with you"; in French "au revoir" means "until we see each other again"; in *Inuktitut* an expression for "goodbye" does not exist at all, although several words exist for greeting new arrivals. No word for goodbye? Peoples who do not speak alike cannot really think alike.

Linguistic scholars have identified no less than eleven different groups of native languages in Canada. Of these fully eight are represented in British Columbia and two—Athapaskan and Eskimo-Aleut—in the Northwest Territories (Figure 1-02). Eskimo-Aleut is the only group of native languages spoken in both the old and the new worlds. Chukotan, a Siberian Inuit language, relates closely to Inuktitut, the Inuit language spoken in several dialects from Alaska to Greenland. Inuktitut has remained remarkably consistent despite this enormous geographical stretch, so much so that the language can be used to discuss issues of mutual interest to all Inuit at

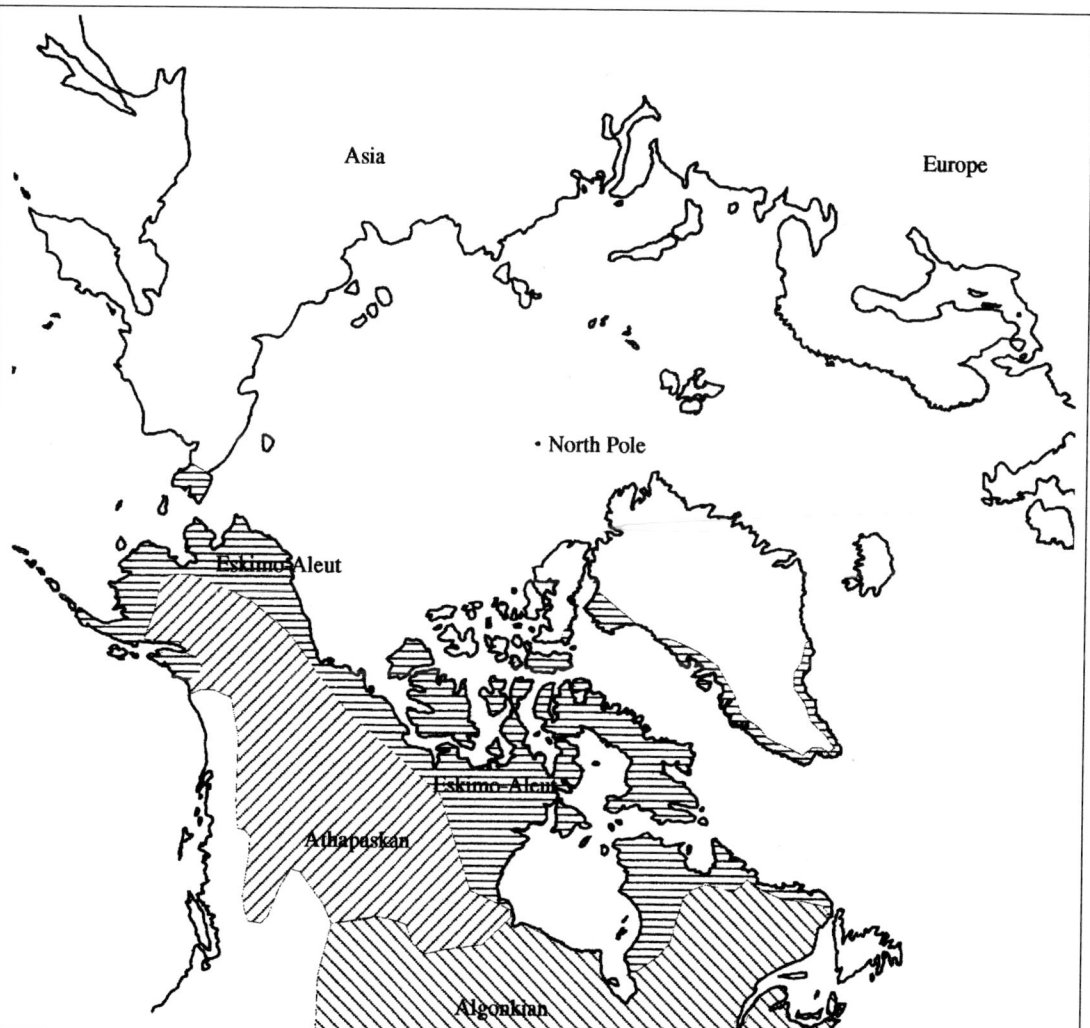

Figure 1-02 The extent of the two major linguistic groups—Athapaskan and Eskimo-Aleut—present in northern North America in the eighteenth century (after *The National Atlas of Canada*, 5/e).

Inuit Circumpolar Conference meetings. It is the lingua franca of the Inuit. The Conference has even raised the possibility of co-ordinating the regional development of Inuktitut (based on Greenlandic, the dialect of the Inuit majority). The advances made by electronics in the field of text preparation and transla-

10   PEOPLE

Planning tomorrow's survey work, Devon Icecap—August

tion, the availability of satellite transmission between isolated communities, and new political will to reverse the language loss of recent decades guarantees the passage of Inuktitut into the next century. At the moment Inuit Canadians use two written forms of Inuktitut, syllabics in central and eastern Canada, Roman orthography elsewhere.

The Athapaskan family of languages has not fared as well in the Northwest Territories. Since speakers of these languages live geographically closer to the non-native, unilingual majority of Canada than do the Inuit, language loss has been greater. Although the losing trend was reversed in the 1980s, a shortage of resources slows the recovery. Linguistic diversity within the group, imperfect knowledge of the written form of Athapaskan languages, and small numbers of speakers complicates the recovery process. Chipewyan, Dogrib, Gwich'in, North Slavey, and South Slavey are not mutually intelligible constituents of Athapaskan. Dogrib speakers must use English to communicate with Gwich'in speakers. English, the very language that threatens the existence of Athapaskan languages in the Northwest Territories, has supplanted sign language as the lingua franca of the Dene/Métis population.

Building designers in the Northwest Territories speak English, and will probably continue to do so until replaced by native designers sometime early in the next century. Few designers speak any of the native languages. The best of the few available translators and interpreters work exclusively for high levels of government, leaving the least-trained to work at the community meetings called to discuss building projects. Native languages do not contain equivalents for many English technical terms, although committees on native language development address the problem periodically. For these reasons alone communication between native users of buildings and non-native designers remains too rudimentary for them to agree on any but the basic requirements of a new building. The subtle requirements rarely survive translation.

3 TRANSITIONS

THE INVASION

Human history has many instances of native societies being steam-rollered by colonizing societies. Whole peoples (the Beothuks in Newfoundland and the Arawaks in the Caribbean to name two) might disappear, overwhelmed, killed off by invaders. Exceptionally, the native society, with some unexpected advantage, might absorb the colonizing one over time through intermarriage, and turn the tables (as the Greeks turned their Roman conquerors into Greeks during the last century BC). In the Americas, in the "New World," the rule, rather than the exception, has held sway for the last five centuries. Strength has been counted in weight of numbers, industrial technology, political ruthlessness, and whatever momentum impelled the "strong" to risk departing their place of origin.

To the colonizing society the native society's evident success in developing and prospering on its own territory for generations is immaterial. It makes no difference that the native society is perfectly attuned to its environment, to its spirits, and to its vision of the future. It makes no difference that the native society is flourishing. Might makes right. In the Americas European might made right in one demographic region after another in less than a hundred years. It decimated aboriginal populations; it laid waste to the aboriginal culture and economy; it humiliated aboriginal survivors by decree. Of an estimated forty million natives living in the Americas in 1500 AD, fewer than a fifth survived. Most of those who perished succumbed to European diseases for which aboriginal society had little time to develop immunity or remedy.

By the middle of the twentieth century the invasion of the Americas was largely complete, although after-effects persisted. Small populations of aboriginals continued to exist on miniature territories (reserves) here and there in the new hinterlands, or in remote barren lands or tropical forests of uncertain value to post-industrial society. Then, almost unaccountably, self-doubt crept into the collective mind of the European majority. Post-industrial society, having won leisure time for the masses, was able to pause long enough to begin putting the considerable cost of its "progress" in perspective.

Self-doubt slowed the steam-roller to a snail's pace, enough to allow aboriginal peoples to begin to regain their poise, to regroup politically, to absorb European techniques for manipulating public opinion, and to state their case decisively for the establishment of self-governing, distinct aboriginal societies living interdependently with the rest of the world. By the 1980s the steam-roller had lost the initiative.

KINSHIP

Aboriginal societies in northern North America were self-reliant. A hundred years ago the Polar Inuit of northwest Greenland (opposite Ellesmere Island) believed that they were the only human beings in existence. The fish and game necessary for subsistence was spread so thinly over such a vast territory (or, at times, in huge concentrations at inconstant locations) that people limited in range by the distance their own feet or the dog team would carry them could subsist only in small groups. There was not enough food and clothing material within range of a five-day walk to support a large group. Self-reliance was an economic necessity rather than a philosophical concept. The band economic unit, typically three to five extended families, took its base and sense of identity from a specific stretch of land. The band knew itself by its attachment to familiar landscapes and distinguished itself from other bands accordingly. It moved from place to place within its territory at the dictate of seasons, of fish and game, reoccupying the best places year after year.

The hunt was paramount. A poor hunt forced the hunters to feed themselves first to enable the hunt to continue. Old people and children ate last in such circumstances. Sharing of the harvest and division of labour were settled in accordance with custom. The instinct for group survival prevailed over the instinct for survival of the individual; band society was a stratagem for group survival. The hard lives of individuals did not have separate importance; the band was bigger than self. Within that limit family relations were very close, and closeness in turn demanded tolerance of different views and unexpected change. Over a disagreement, or for any reason at all, a family or families might leave the band to join or form another band, never to return, or to return months or years later. In order to defeat high infant mortality, many children were born. Childless relatives, especially the older people, adopted surplus children born within the band and in neighbouring related bands. Kinship ties were extensive and complicated, the glue that sealed peace and co-operation between related bands in disparate locations.

Unrelated bands often fought each other for booty and for female slaves. Fear of the surprise encounter and revenge for past atrocities fuelled enmity. Despite the vast distances separating the two peoples in most regions the Dene and the Inuit also skirmished, with deadly results.

Band leadership was not formal in structure or duration. The hunter who spoke for the most people, with the most sense, most often had the final word. Usually an elder, he understood best not just the habits of animals on the land but also life's lessons as passed down in storytelling, dance, and games. Sometimes he was the shaman. In a society without books, the able older person, the longest-living "book," kept the respect of his people.

CONTACT

In North America Europeans made the first recorded contact with aboriginal societies just after 1000 AD. Norse people sailing west from bases in south Greenland explored the rough edges of the North American continent, reaching perhaps as far south as the mouth of the Saint Lawrence River and perhaps as far north as Ellesmere Island. An attempt by the Norse to settle in Newfoundland appears to have ended in failure almost right away, frustrated by skirmishes with aboriginals bent on repelling invaders. Wherever they found aboriginals, deaths on both sides soon followed. Thereafter the Norse may have continued to visit those shores only long enough to cut logs, a commodity plentiful in Labrador compared to Greenland. These glancing contacts quickly became legend, but Norse artifacts did make a material impression on North America. The Thule people, predecessors of the Inuit, had no previous knowledge of iron and its uses, but archaeological evidence suggests that the Thule used iron tools made from ore originating in Greenland within a few generations of the first contact in locations as far west as the central Canadian Arctic. Similarly, but hundreds of years later, European iron found its way from the Pacific coast of North America to the Athapaskans living in the western Canadian subarctic. The first European traders in northern Canada did not venture into the interior but remained by the sea in Hudson Bay and along Canada's west coast. Aboriginal middlemen carried their tradeware into the interior. Since iron implements substantially reduced the uncertainty of subsistence in northern Canada, a sellers' market for European technology developed almost overnight.

As overtrapping depleted the supply of fox and beaver pelts near the coasts, and competition between European trading companies for the dwindling resource increased, the traders decided to resettle closer to those aboriginal people willing to do the work of tending traplines. European presence became permanent; by the early nineteenth century a network of forts and settlements had been pushed into the northern interior.

Europeans and Americans also set up arctic whaling stations, enlisting Inuit help to harvest (and render nearly extinct) the whale populations of Davis Strait, Hudson Bay, and the Beaufort Sea. The impact on native societies was immediate and irreversible. Infectious disease, hardly controllable in Europe, ravaged New World populations whose existence separate from the Old World for thousands of years meant zero natural immunity. As recently as the beginning of the twentieth century the Mackenzie Inuit disappeared from the Mackenzie River Delta, killed by diseases brought by outsiders. (Easter Island, an isolated microdot on the map of the South Pacific, built its first airport in the late 1960s; its population had to be specially immunized against the potential for disease represented by the first wave of tourists.) Similarly, low tolerance for alcoholic beverages imported by Europeans made alcoholism an endemic disease.

Traditional aboriginal diet did not include fox, beaver, and rum. Band society had to reorient itself to the fur trade and the whale trade. Pelts and labour were exchanged at the European outposts for flour, sugar, tea, tools, clothing, rifles, and ammunition. The benefits of this new way of life seemed obvious. Guns meant that the hunt was simpler, surer, if not less time-consuming; guns meant that more dogs could be kept and fed—enough dogs to make a team suitable for extending the range of the hunt. Since most of the items received in trade from the Europeans could be consumed in a month, the pressure to replenish them became intense. Replenishment meant that more fur, and more labour, had to be offered to the traders.

Aboriginal populations, hunters for millennia, became part-time trappers. If one half of waking time was spent on trapping and the other on hunting for meat, then only half as much food would be taken from the land. The other half had to come from the traders' stores in exchange for fur. When the trading post closed (posts might remain open for two years, five, or fifty) on instructions from headquarters in London or Montreal, its native clients had suddenly to fend for themselves again. Or if a new post opened three hundred kilometres away the burden of transporting furs to the new location fell to the trappers. Band society lifestyle hung in the balance. The traditional safety net, self-reliance, had been left to decay.

DEPENDENCE

Over the last two hundred years aboriginal dependence on the presence of settled Europeans has had a corrosive effect on traditional lifestyles. Band society understood that human effort alone might not be enough to prevent human misery. Skilled hunters could misjudge the path of the caribou migration and fail to make contact with the main herd—starvation would ensue and all or some of the band might perish. To the Canadian government, apprised of death by starvation in the south Keewatin barren lands in the 1950s, the risk was unacceptable. The government determined (without consulting those affected) to extend welfare to the peoples of the north. To respond to the primary needs of tiny populations spread out over vast stretches of land, a cheap, practical method of delivering basic services had to be devised. Since the cost of extending a services pipeline from Ottawa to each of a thousand moving dots on the tundra would be prohibitive, people on the land were obliged to live in designated centres—settlements—where the services could be more conveniently delivered. The problem was shaped to fit the solution.

Contact with settled Euro-Canadians meant exposure to European ways of thinking. Euro-Canadians brought the beginning of North American history with them; the millennia spent by aboriginals in the arctic and subarctic were called "prehistory." Conventional wisdom relegated the oral history tradition maintained by band society to second-class status. The Euro-Canadians, by "discovering" the arctic, arbitrarily set the clock back to zero, devaluing native culture in the same instant. To European culture this was a natural perspective devoid of premeditation.

The evangelization of entire native populations was certainly premeditated. European missionaries admonished natives to abandon their own metaphysics in favour of a spiritual world and ethic devised by foreigners. First the missionaries, then the Canadian government, set up residential schools in which children from outlying settlements (the majority) were taught a foreign language and culture at the expense of knowing their own. The physical separation of children from their relatives and from the landscapes they were born in caused deep distress and a generation gap that has widened into a permanent rift. The physical and moral abuse heaped on many of these wards of the state by some of their teachers opened wounds in native society that have not healed to this day.

The matter of residential schools points up the fact that, like other societies, native societies do not respond to situations and ideas monolithically and reflexively. While for some the history of the residential schools and the family disintegration they caused are too painful to contemplate, others believe that the aptitude for study fostered by a strict school regimen helps to build the kind of leadership resource that native society as a whole sorely needs for the challenging times ahead. The former want to slam the door on the remaining residential schools; the latter want to open residential schools for students with leadership potential.

Later, the government opened elementary schools in the settlements to reduce the incidence of physical separation. Still, children following a nine-

Cambridge Bay—March

Repulse Bay—March

to-three school schedule could not at the same time be on the land learning the ways of the hunt from their kin. Housing erected by the government to serve native populations was designed on a pattern suited to southern Canadian suburbs. The houses, unprepared for the arctic and subarctic, were small, cold, and drafty. The first houses built and most of their successors failed to accommodate aboriginal lifestyle: native people had to adapt their lifestyle to the house. In southern Canada a house design unadapted to the lifestyle of potential clients would disappear from the market within a year for want of a buyer. In northern Canada the potential client had no choice.

The Canadian government made elementary health services available to each settlement through a nursing station. People with difficult health problems were evacuated to medical centres in southern Canada. Even giving birth meant sending the mother away to spend weeks or months in a totally foreign world far from family. In the 1950s tuberculosis patients were sent out of the settlements to sanatoria in southern Canada. The result, usually a cure, meant loss of contact between patient and relatives. Health officials, confused by inconsistent spelling of native names and complex kinship ties, failed to identify the relationship between individual patients and their healthy relatives prior to evacuation and the patient became "lost." The youngest patients knew neither their parents nor where they lived, and so could never rejoin them; instead they joined the urban poor in southern Canada and were completely cut off from their roots. Many of the patients who did die were buried in southern Canada at locations unknown to their relatives.

Adult members of a band subsisting on the land were always fully employed. Life in the settlement imposed the concept and the fact of unemployment. A hunter living in a settlement has no livelihood. There are few jobs, and less training. Seminomadic people used to being perfectly fit become sedentary: depression, obesity, and diseases of the heart result. Suicide becomes commonplace. The food bought at the trader's counter is expensive and nutritionally poor. The baby teeth of children decay at an unprecedented rate.

Natives new to settlement life had to learn the clock: wage labour time has a start, an end, fixed meal times, and coffee breaks. Many things are done when it is time to do them rather than when it makes sense to do them. The band lifestyle does not function by the clock. Some settlements encountered the bust and boom economy of resource development—mining for nickel, gold, lead, or uranium—causing another new kind of stress: for five years there are fifty jobs, then there are none. These settlements cannot become ghost towns like those in southern Canada, because the settlers have no other place to call home.

Natives in northern Canada's remote settlements rarely leave home in search of opportunity outside. The stress of separation from kin, lifestyle, and landscape outweighs the supposed rewards of making a new life in a strange place. In comparison, the Canadian population as a whole is extremely mobile—many are quick to pack up and leave to exploit the latest economic boom no matter how far away. This essential difference in mobility provides a prime example of the inability of cultures in collision to recognize and understand each other's deepest motives.

DIRECTIONS

Northern Canada is a Third World country in everything but name. It is underdeveloped. It is rich in non-renewable resources but unable to exploit them because of high development costs and low international commodity prices. It has a system of highways inadequate to the task of shipping goods to markets. It has native culture in collision with imported culture. It has an immature political system. It has high unemployment and lacks a stable pool of entrepreneurial and technical skills. It expects to become three territories in the place of two. It is short of capital investment. It faces an uncertain future.

The means by which one culture can successfully direct and administer another have not yet been discovered. Only self-government, making people responsible for the conduct of their own affairs, resolves the impasse. Northern Canada, and other circumpolar regions, are making progress toward this end. (Greenland, by exercising Home Rule, has made the biggest step to date.) At the same time the outside power has to step back, fully prepared to co-operate under new rules of engagement.

In northern Canada the progress toward native self-government comes in the form of devolution of power from the federal government in favour of territorial governments, and final agreements between native groups and the federal government on land claims. The first gives territorial governments more and more political power until they finally resemble Canadian provincial jurisdictions. The second gives native groups economic power in the form of starting capital and a large land base. Settlement of land claims seems to require in excess of ten years to complete: there are several regional agreements in various stages of negotiation. The protracted delay in ratifying agreements has caused economic uncertainty for both insiders and outsiders. But once ratified the agreements will permit native-run corporations to set the ground rules

Upernavik—August

for future development of resources and to invest in existing business or in joint or wholly-owned ventures.

There are many lesser agents of change. Co-operative societies formed to sell native crafts and art to the public at large while providing goods and accommodation services within settlements prosper. Tourism associations promote destinations in each of the regions of northern Canada. Cultural institutes exist to advocate preservation and extension of Dene and Inuit language and culture. Federal and territorial land use planning and town planning initiatives now stipulate native participation. Broadcasting societies exist in many communities; native programming, fed to all of northern Canada by geostationary satellite, steadily develops its potential as a unifying and co-ordinating force. Wildlife management, the management of a renewable resource, has progressed from cross-purposed jurisdictional battles between native hunting associations and government agencies to competent administrations committed to sustainable harvest.

European and American anti-trapping lobbies have pushed the trapline economy to the point of extinction: full recovery is an open question. Northern Canada has never been able to grasp the fact that city dwellers in western Europe could demolish a rural way of life in a remote part of the world without a second thought. More insidious still, the ability of industrial societies at midlatitudes to introduce pollutants into the northern environment eludes comprehension. High residual levels of PCBs and mercury found regularly in animals and fish harvested for food by northern residents, and arctic haze and depleted ozone in the high atmosphere, are just two aspects of the environmental degradation caused by midlatitude industrial society.

Progress toward native self-government includes the replacement of transient outsiders in public and

Discussing the new school project—Sanikiluaq

private administrations by trained, able native residents. The shortage of such replacements presents a difficult problem. If education and technical training truly aims to broaden horizons and such broadening cannot be achieved realistically by teachers working alone in isolated communities, then there has to be a greater readiness in individual trainees to leave the settlement for periods of study in larger centres. Like learning to live on the land, education is in part learning by example, accumulating experience good and bad: book learning by itself cannot accomplish the levels of training needed to direct society wisely. Although politicians try to bend the existing education system in northern Canada (which is based on the provincial models of southern Canada and largely staffed by outsiders) into something that begins to serve native people properly, it seems likely that a complete renewal will be necessary.

Evolution of the justice system suffers similar impediments. The abolition of the native justice system and the application of a foreign one have resulted in a messy transition, a transition that will take at least another generation to complete its course. The bureaucratic delay between the commission of an illegal act and legal retribution combined with the continued, unrestrained presence of the accused in the home community also causes hurt and confusion among victims and residents alike. This kind of impasse has led some native residents to advocate a legal system that obliges the wrongdoer to make restitution to his victim. Making amends would replace taking revenge.

Another generation at least will be required to right the situation in the field of building and community design. There are few native building professionals. The majority of building professionals in northern Canada are transients with incoherent and incomplete ideas of native reality. In the short term these people will be replaced by natives with equally incoherent ideas about the realities of Euro-Canadian life and native society in transition. But in the phase to follow, in this discipline as in others, one can expect that native professionals will make mature design decisions for people whose culture is largely redefined and understood.

Judging from current events in northern Canada and in other parts of the Third World, native culture in the twenty-first century will likely be an extension of the conventional wisdom, taking the best of both worlds. Although vigorous, the 1970s back-to-the-land movement did not slow the changes underway in mainstream native society; in material terms it only led to bringing the twentieth-century goods and services network to ever smaller, ever more remote settlements, at ever greater cost to the population at large. Physical isolation from the negative aspects of life in larger communities has not resolved the complex issues of a society in transition. But the resolute spirit of such self-imposed isolation underscores the depth of the growing determination by native residents to do much better as a society than standing transfixed in the path of the steam-roller.

4 LIFESTYLES

SNAPSHOTS

Industrial society, none too careful where it treads, perceives polar regions as frozen landscapes sparsely peopled by rugged individuals coping fatalistically with impossible odds. Humans versus the environment is the standard theme of treatise, reportage, novel, and movie. The reality, what local people do and think ordinarily, staggering beneath the weight of romantic myth, barely makes itself felt beyond the regions' boundaries. The myths, while interesting in themselves, do little to inform the people responsible for high latitude building design about the ordinary needs and concerns of the people that live there.

By means of the "snapshots" that follow, this chapter steps around the myths and exposes just a little of the ordinariness that makes up everyday life in polar regions. (Given names and some place names have been altered.)

Kugluktuk

Joshua, white-haired and in his seventies, runs a no-nonsense video camera in a television studio. A mission school, hundreds of kilometres to the west of his parents' hunting grounds, hammered Inuktitut out of him when he was a boy. Much later he relearned his mother tongue to join the handful that still speak the full vocabulary with knowledge of old traditions and beliefs. Asked to delve into those beliefs, most elders fall silent, concerned that even today's church leaders might disapprove. Joshua can sometimes penetrate that silence by evoking the concern common to most elders that the values inherent in the old way of life are being forgotten like cast-off clothing, displaced by an array of disconnected ideas that, in younger people, seem to end in anti-social behaviour. Sometimes he thinks that preservation of the old values and the language is a lost cause. But with so much energy left, and retirement out of the question, Joshua continues to promote the use of Inuktitut and to record the stories of his contemporaries.

Cambridge Bay

The old school here, with all its blind corners, counts as an obstacle to ATV (all-terrain vehicle) traffic. Today five tradesmen are doing repairs on its roof. They unroll rolls of new waterproofing membrane at the roof's edge ten metres above the ground; two hold torches for welding the joints of the material. They notice three young women walking into view around a corner of the building. The roofers also see a four-wheeler approaching quickly on a collision course from the other direction. The women, accustomed to four-wheel eccentricity, remain unconcerned at first. But in an instant two of the women jump out of harm's way. The woman in the centre freezes to the spot.

Finally braking, but too late, the rider drives his machine into the woman, propelling her five metres backward. The ATV engine dies. For three seconds time stops. The woman lies sprawled on the ground, semi-conscious, partially unclothed by the force of the blow. The bystanders on the ground are in shock; the roofers above are in shock. At last one of the victim's companions reaches out to draw down the errant blouse. Time starts again. Articulation made difficult by righteous indignation a roofer screams down at the rider, "Are you crazy?" The driver, without a pause, on the offensive, screams back at the roofer, "What's your problem, pal?" as though to confirm that life has indeed returned to normal. Net result for the victim: a broken collar bone, skin abrasions, mental trauma; for the perpetrator: a $75 fine.

Repulse Bay

Christine, regal in her sixties, wears a bright pink top over flame-red pants. Her high boots are made of soft sealskin prettily embroidered. She holds court daily at her chrome-and-formica kitchen table, in a community peopled by her children, her grandchildren, and their cousins. She has many stories to tell. As a child she learned to carve by imitating her father. She has worked soapstone, whale bone, walrus tusk, antler, and even wood. Soapstone no longer, because of the sickening dust. People still press her to carve objects for themselves but she relents only when the clamour is insistent. She keeps none of her work.

Paulatuk

The old community hall has been set up to receive the circuit court. At the far end, at separate tables, sit the judge and the two opposing counsel, dressed in city clothes covered by flimsy black robes. A court stenographer places herself strategically to hear everyone. The vitally interested members of the public sit on stackable chairs near the middle of the room. The mildly curious sit on benches fixed to the side walls at the end of the room where everyone enters but the judge. People shuffle in and out during the proceedings and heads keep turning to monitor arrivals and departures. The little ones play as best they can in a tense, stuffy atmosphere. The only native present at the formal end of the courtroom is sitting in the dock accused of sexual assault. No one disagrees about the facts of the case. The issue is punishment, where and how long. The victims are dead tired of having the accused live among them months after the crime, free to threaten and taunt them as though nothing had happened. On a technicality the decision is postponed.

Fort Providence
On the soft shore of the river a distinct set of tracks works like a signature. In a moment the outboard is stilled and the motorboat glides ashore. Two men armed with rifles step out purposefully, one to enter the woods to flush out the signatory and the other to wait in ambush farther down the shoreline. A third man, an eight-year old, has instructions to remain with the boat. Soon an unaccompanied moose appears between the woods and the shore line, a hundred metres from the boat. No sign of the hunters. The third man, the boy, lifts a hunting rifle from the bottom of the boat, sights the moose and pulls the trigger. Nothing. The chamber is empty. But there is still time. The second pull sends a bullet into the chest of the moose. A few moments later the chest of the third man's father bursts with pride, and the third man himself, losing his exquisite cool, begins to tremble.

Hay River
The expression, "Hello. My car seems to have an electrical problem. Could you look at it?" elicits a stony stare from most garage mechanics. Perhaps they are hardened by the belief that they are treated as saviours by the mechanically inept when the car breaks down and as lowlifes the rest of the time. Mike even has the reputation for being downright rude to strangers, but he does get the job done. Many years ago he turned up at a lumber camp in the Peace River country direct from Italy. He spoke no English. Nor did the Cree lumberjacks. The first Cree words Mike learned are unspeakable. Twelve years later he went north to work for the Northwest Territorial government as a heavy equipment mechanic. By then he had a Cree wife, who soon after left him, and five young children who did not. He became a single parent. He sustained workplace accidents, including the loss of a foot. He taught mechanics at the trade school until his students had had enough of his blunt manner. He then found independence in a self-owned garage, remarried, and graduated from grade twelve. His second wife lives in British Columbia. His youngest son works beside him in the garage.

Taloyoak
The beaten patch of snow behind the house serves Joseph as a carving studio. Like homing torpedoes, the soapstone sculptures he carves search out wealthy buyers thousands of kilometres away. Today he applies a rotary file to the rough parts of a ten-kilogram shaman through a cloud of rock dust. He works outside so that the dust does not infest his lungs and the lungs of his wife and his daughter's family. There is sunlight but it is still winter, and every half hour or so the machine noise stops and Joseph goes through the back door into the kitchen, where a tea kettle slow-boils. Though new, the back door already shows the scars of heavy use. The door does not close properly, because the edges around the lower third are covered with condensation ice. The building designers failed to anticipate that the back door to the house would be the front door to the studio. The door installed was intended for occasional use in winter. The designers have never met Joseph, who has been carving soapstone and ivory for thirty years.

Lutselk'e
Living in the big city among strangers far from home is tough. Homesickness is endemic. But going home to the village much later is tough too. Relatives and old friends are suspicious of the person made sleek by city ways. There is a proving time; acceptance has to be earned. The city "I" has to be replaced by the village "we." Sarah, slight, sharp-featured, shy and humorous by turns, has been through the mill. In Regina she thought she wanted to be a biologist, but looked into training to be a nurse. Discouraged by the length and the loneliness of the path ahead she returned to the village and fished commercially for a while before becoming receptionist, then secretary, then acting Band manager and accountant at the Band office. The shifting nature of Band and government politics tended to imperil the continuity and completion of her community projects, so Sarah happily accepted work first in the office of a local construction company and then, braced with some formal business training, in the office of the local economic development corporation as manager. Now she feels in control.

Iqaluit
There is a severe housing shortage in the North. Some people live in tents, waiting for houses to be built. Some live in the furnace rooms of houses allo-

cated to relatives. Keekoogak's son, his son's wife, and their four youngsters sleep on the linoleum floor in the living room. Mattresses are laid down every night. Children seem to be everywhere. Keeping the place tidy is not easy when twelve people live together in a two-bedroom rental unit. The water vapour produced by twelve pairs of lungs, the extra washing, laundry, and tea boiling makes the *relative humidity* soar. For lack of curtain rods that would leave an air space between curtain and window frame, patches of cloth have been tacked directly to the window frame, cutting off air circulation against the window glass. Without the warming effect of room air the window glass temperature dips below *dew point* and the windows ice up. The ice remains for the rest of the winter, until spring sunlight starts a flood of meltwater.

Tsiigehtchic

Two outsiders, men in their sixties, look-alikes in physique and vocation, work at nurturing the memory and teachings of Jesus in a white clapboard church. Inside two separate traditions prevail. When services are held on both sides at once, the singers compete by raising their voices. In summer these glorious sounds drift down to the big river, a quarter kilometre away, and lap against muskrat eardrums.

Rankin Inlet

At latitude 62, five degrees south of the Arctic Circle, the sun does not disappear altogether in late December. It sports a pale impression of itself for a few hours around noon. By early March the sunlight is stronger, intensified by reflections off the snow, and stays longer every day. In the lee of the wind March sunlight feels warm to the touch, but the winter norwester blows for another four months. Finding a spot outdoors that combines sunlight, windbreak, and a place to sit is next to impossible in this community. Nowhere for people to sun themselves like seals on the spring pack ice.

At Nuuk, West Greenland, a few such sunny spots have been built purely by accident. One spot consists of the flat roof of a one-storey wing abutting the high south-facing side wall of an office building. For those teenagers willing to make the three-metre climb up to the roof (the majority), this sunny "balcony" provides a place to loaf while looking down on the busiest part of the downtown. Shielded from the wind, exposed to the sun, this roof has become an important hangout, a palpable influence on the way people grow up in this capital city.

At Rankin Inlet the student hostel offers a sunny niche by design. A balcony, an extension of the main entrance landing, lies in the lee of the prevailing wind, and faces the sun at the equator. In fall and spring low-angled sunlight floods the space, which is provided with wooden benches around the perimeter. Each year new generations of students adopt the place, clearing it of snow in the spring to make space for sitting. Even at freezing air temperatures the niche offers them a place to rest or read next to a point of interest (the main entrance), a catchment for direct and reflected sunlight, a windbreak, and a seat that warms up on contact with a live body.

Pelly Bay

The community centre building sits on exposed bedrock about five metres above the rest of the settlement. It fits neatly into its setting. The main entranceway lies at the top of a gravel slope more suited to pedestrians than to vehicles. There is no step. Inside, one flight down, there are hamlet

High-rise log cabin. Whitehorse—October

offices. The Hamlet Council and the local education authority have gathered in the council chamber one flight up to hear recommendations from a transient flock of civil servants regarding siting proposals for the new school. Sunlight floods into the chamber and the adjacent meeting room through skylights. From the meeting room the sunlight spills one step farther through a glazed partition into the upper space of the gymnasium. The echoes of desultory hockey play rise from the gym floor to the council chamber. The council meeting gets

Modern cliff dwelling. Yellowknife—April

underway when its senior member is made comfortable. He arrives at the main entrance in the passenger seat of a pickup truck. A hulking man fifty years his junior carries him the remaining distance through the lobby and up the stairs. The senior councillor has not walked for twenty years. He leaves his wheelchair at home, because the only spaces in the community centre accessible to it are the lobby and the gymnasium.

The nursing station across the road has an enormous wheelchair ramp made of steel outside its main entrance, but the toe of the ramp has never been installed. The ramp ends in a one metre free fall.

Sanikiluaq

This is the fifth new house to be inspected today. One purpose among several is to examine condensation ice on windows and water staining on and below the sills. On other visits this spring to new houses in communities farther north, water damage from melting ice around windows has been severe. After an exchange of greetings the household falls silent as the visitors make their rounds. Silent except for Elizabeth, aged three, dark and beautiful in pigtails. Self-possessed, full of intelligent curiosity, and plainly used to dominating the society around her, she begins to sway the visitors as well, watching their movements intently, moving with them or ahead of them, exclaiming occasionally, her retinas evidently taking in every snippet of idiosyncratic behaviour.

The telltale signs of water damage at the living room windows are not there; in the other rooms the signs are slight. Why should this house be different from the others? Why is there no water damage? Elizabeth's grandmother replies that she routinely attacks ice formation on windows with a hair dryer and wipes away the meltwater. She never lets the ice get ahead of her. (If everyone looked after their windows this well, building designers could continue to detail windows badly.) In parting, a visitor asks permission to take Elizabeth away to a new life in the big city. Before her mother can say "No!" a stricken look flashes across Elizabeth's face, to be replaced just as quickly by a look first of relief and then of feigned nonchalance. Elizabeth cannot be fooled.

Yellowknife

The city's survey plan shows a house lot here, but in reality there is nothing but a huge Precambrian outcrop falling steeply down to road level, where a space the size of a living room harbours a few birch trees. The drop from top to bottom must be fifteen metres, and the slope varies from thirty to ninety degrees. What is such a tight-fisted site good for? There is no room for a house. The most that can be built here is a stairway to the top to get a view of the big lake, just like the stairs that lead up to the Pilots' Monument perched on the outcrop at the other end of the causeway.

Stairways it is. The house here has two. The first, beginning at road level, open, made of wood and married to the cliff face, snakes up to a landing at tree top level. There a main entrance leads to the second staircase, this one totally enclosed in a wooden shell. Straight and parallel to the rock slope, here about thirty degrees, the second stair ends with a back entrance to a wood deck patio at the top of the outcrop. For every three metres of rise there is a landing.

For every landing there is a clutch of rooms, to the left and to the right, all enclosed in the shell, but most without partitions. One can see and shout from top to bottom and be heard. At intermediate landings there are skylights in the roof, some hanging plants, and a few windows. A gang of windows at the top gives a spectacular view of the lake and the floatplane base. The second stairway forms the spine of a modern northern house. It is the spine of an oblique parallelepiped whose walls, without exception, are vertical, and whose underside, ceiling, and roof, without exception, are sloped. The parallelepiped floats above the outcrop on wooden stilts. The survey plan was right. There is a house lot here.

II
TERRAIN

5 EARTH AND WATER
SPRING MELT *23*
ICECAP *23*
GLACIER *23*
ICE *24*
SNOW *25*
PERMAFROST *25*
EARTH *26*
PATTERNED GROUND *26*
HUMAN TRACES *27*

6 AIR AND FIRE
AIR OCEAN *28*
SOUND *29*
AIR QUALITY *29*
OPTICAL PHENOMENA *29*
NORTHERN LIGHTS *30*
VOLCANISM *31*
FIRE *31*

7 FLORA AND FAUNA
DESCENT TO THE OASIS *33*
THE TUNDRA *34*
THE TAIGA *35*
THE OWL AND THE RAVEN *36*
OTHER ANIMALS *37*

Virginia Falls. Nahanni River—July

22 TERRAIN

*Glaciers on the move.
Axel Heiberg Island—July*

5 EARTH AND WATER

SPRING MELT

From the edge of the Devon Icecap the Sverdrup Glacier flows northward twenty kilometres through a U-shaped groove two kilometres wide—there are four doglegs—and empties into Jones Sound near 76 degrees north latitude. At its high end the ice surface is 1,200 metres above sea level, very white, smooth, and regal, but at the snout where the glacier calves noisily into the sea the surface is dirty, wrinkled, and crevassed. From the snout a view opens northward onto the dark cliffs of southern Ellesmere Island seventy kilometres across the sound. Invisible at the base of those cliffs lies Grise Fiord, Canada's northernmost settlement.

By the end of June, after two months of around-the-clock sunshine, the only signs of melt are occasional rock falls at the valley walls and the snow surface turning mushy by noon. At night, under the cool midnight sun, the snow surface becomes crisp again. Only two, we cannot manhaul the loaded sled under the noonday sun—the sunlight overheats us, the wet snow offers too much sliding resistance, and our patience expires quickly. We decide to do our glaciological work at night, when bodies, minds, and snow surfaces are crispest. We sleep during the day.

After a week of this, moving first down and then up the glacier, exchanging a night of labour for a day of fitful sleep, pitching our tent wherever the work stops, we decide this morning to make camp on the medial moraine, that strip of fallen rock strung down the middle of the glacier all the way to the sea like a set of vertebrae. We find a flat spot among the boulders just wide enough for a tent. This vantage point offers a spectacular view and a sense of being at the centre of something vast, quiet, and more indelible than the puerile chatter coursing through our minds. The heat of the noonday and every photon of light reflected off the surrounding snow engulfs the tent until the somnolent brains inside seem to boil over. Then, without warning, a noise outside the tent breaks our reverie—not the familiar, discrete rumble of a rock fall but a muffled roar increasing in intensity. One head pokes through the tent flaps. There is an investigatory pause, soon followed by incoherent shouts and two half-dressed bodies bursting from the tent.

The roar precedes a wall of slush half a metre high by ten wide coursing down the glacier faster than a person can run. The front swallows all the snow in its path. In seconds it passes within a few paces of the tent and crosses the medial moraine a hundred metres downstream. In minutes the stream of freezing water behind the front accelerates to bobsled speed, opening a channel three metres deep in the blue ice of the glacier. In hours the meltstream re-establishes itself in last year's channel, hidden until now by the winter's snowfall. And in days a glittering field of bare ice crisscrossed by tributaries replaces the blanket of snow covering the glacier. A glance at the pace of the stream in the channel confirms the obvious. Anyone falling in would be quick-frozen and swept away. The channel is icy smooth. There are no hand holds. Mesmerized, we feel lucky to have been watching from the safety of the moraine. And with the soft snow gone from underfoot we can go back to working in the daytime.

ICECAP

Like starlight, California redwoods, and raised beaches, the icecap is a time machine. Each snowfall records the passage of seasons. Core samples of ice, from holes drilled more than a kilometre into the surface, reveal annual layer lines (similar to tree rings) that contain trapped gases, pollen, sea salts, soils, and meteoritic dust. The ratio of nitrogen, oxygen, and carbon dioxide in the atmosphere and other climatic parameters of the year 1066 AD—the year that William the Conqueror defeated Harold at the Battle of Hastings—can be determined with assurance. Icecaps will eventually reveal human artifacts such as World War II fighter-bombers, toothpaste, and harmonicas. Even buildings poorly adapted to the icecap environment may disappear under snowdrifts in winter and fill up with meltwater in summer. The icecap swallows whole any stationary object and spits it out at the periphery, crushed, hundreds of years later.

Wildlife on the icecap is scarce. Bird sightings are rare and mammal tracks nil. Occasionally the wind blows an insect this way to die of exhaustion and cold. The closest life—lichen, some arctic poppies, and a few minuscule clumps of purple saxifrage—exists on nearby *nunatak*s, barren exposures of bedrock poking up through the icecap like islands in the sea. Life really begins on the plateau at the edge of the icecap, where valleys strewn with boulders carry ice and meltwater to the sea.

GLACIER

Like any other mineral near its melting point, glacier ice flows under the pressure of its own weight. It is plastic, not rigid. Plant a set of stakes straight across a glacier from one valley wall to the other and not

Stranded iceberg. Qeqertarsuaq—August

only will the whole line of stakes move downstream in summer and fall but the line itself will bow downstream. The stakes in the centre move faster than the stakes at the sides. All the stakes at the surface move faster than the ice near the bottom of the glacier. For part of the year polar glaciers may shift as much as several metres a week.

Crevasses in the ice open and close at changes of slope in the valley bottom, at doglegs in the valley walls, and at the snout, where the glacier stumbles, disintegrating in sudden stages to make icebergs. Summer heat melts the snow cover and turns the glacier surface into a field of bare ice spotted with ponds of slush or clear water and lined with meltwater streams. At intervals the bedrock forming the valley walls, fissured by thin wedges of ice, sends loose boulders crashing down as the ice holding things together thaws. Some boulders tumble onto the glacier surface to be transported seaward as *moraine*. Dark-coloured rock lying on the glacier absorbs more radiation from the sun than the surrounding ice. When rock temperature rises above water's freezing point the rock melts slowly into the ice surface. Even pebbles disappear down self-made melt holes, making delicate birdshot patterns on the glacier.

ICE

Most of Canada's land mass has been glaciated at least once, and five percent of it remains under ice. Although a small proportion of the world's total water, glaciers and ice sheets represent a large proportion of the world's total fresh water. Ice volume figures heavily in the arguments and counterarguments over the extent of global warming induced by changes in the atmosphere for which humans are responsible. The conversion of ice sheets to water would flood coastlines and dislocate low latitude agriculture on a grand scale.

Seen from outer space, pack ice, the sea ice at or near the poles, is the most significant variable feature of the earth's surface. The areal extent of pack ice affects the amounts of solar radiation absorbed and reflected by the polar seas. The energy balance at the poles in turn affects world climatic change. Sea ice newly formed is "black," its crystal alignment so regular that the resulting solid is transparent. Old ice—multi-year ice—having survived more than one freeze-thaw cycle, has a crystalline structure that is hard, irregular, and translucent. *Fast ice* freezes solid to coastal shallows, forming a seasonal embankment parallel to the seashore against which the tide raises and lowers pack ice. An occasional high tide will force sea water up through the gap to flood the sea ice surface and freeze. Where the shoreline profile drops abruptly the fast ice forms a band a few metres wide. Where the shoreline slopes gently the fast ice is much wider. The constant grinding of pack ice against fast ice raises a ridge of broken ice that may be steep enough to hinder travellers headed out to the sea ice.

Sea ice does not form everywhere. Still poorly understood, *polynyas* are areas of open water where sea ice might be expected to form. A large polynya exists in northern Baffin Bay; a smaller one exists in Wager Bay. A combination of sea currents and water temperature fluctuations keep polynyas open throughout the winter.

Currents and wind shift sea ice constantly. Large leads may open up or close in a matter of hours. A huge ridge of ice may form where ice shifted in one direction crashes into ice held up by inertia or the shape of the shoreline. Leads and ridges block ordinary travel. A traveller on foot cannot see far enough ahead to scout the shortest route around these obstacles.

When the watershed thaws in spring, river ice breaks up and rides a surge of meltwater to the sea.

EARTH AND WATER

Ice floes undercut the river banks or snag on sand bars to pile up in dams that cause floods or divert water, silt, and debris into new channels. The Mackenzie River undercuts Aklavik in this way every spring, and occasionally inundates it. Rivers such as the Mackenzie and the Coppermine that flow toward the poles break up in geographic stages from source to mouth. The snow melt, the spring run-off near the source, is held back by ice dams in the still-frozen lower reaches of the river. The sight and sound of spring break-up on the river, floe grinding against floe and candle ice clinking in a swirling current, draw wide-eyed children to the river bank like a magnet. The rivers become impassable for several weeks. At freeze-up in the fall the rivers are impassable again until the new ice is thick enough to support traffic. These seasonal interruptions jostle every local economy that depends on year-round truck transport.

Ice is larger in volume—less dense, more buoyant—than the water that spawned it. This expansion shatters bedrock. Like a liquid wedge, free water pours into crevices, freezes, and expands, widening the crack but adhering the parts together until spring thaw. When the ice in the crevices melts, the rock is released. The meltwater seeps still farther into the crevices and refreezes when the ambient temperature drops again. This incremental action continues in a given location until a rock tumbles downslope, starting a rock fall. This separation exposes a new rock face to the freeze-thaw cycle.

SNOW

In autumn the failing light and the drizzle change the light and shadow pattern of the summer plateau into a landscape so sombre that the eight months of snow and ice about to come seem welcome by comparison.

First snow highlights relief. North Baffin Island—August

Soaked by the rain, the rocky hills go dark, matching the gloom in the depressions. On occasion, however, the first winter snowfall is tentative, less emphatic than a blizzard. Then the pattern of light and shade briefly returns—reversed this time. The tops of boulders and rocky promontories, swept clear of snow by the wind, remain dark grey or brown, while the depressions and crevices between them turn white with snow. The shocking contrast unfolds intricately, showing up every detail of the filigreed topography like an aerial photograph. For a day or two it banishes uniformity, making the landscape seem intimate, knowable, personal, until the first blizzard buries the scene under a monochrome blanket.

At continental high latitudes the distribution of snow on the ground matters more than the total amount of snow to fall. The prevailing wind moves and removes the fallen snow constantly, crystal by crystal, filling depressions in the landscape and scouring the bumps. Since still air is a good thermal insulator and fallen snow is full of still air, the thickness of the snow cover affects the ground temperature regime. Where snow cover is thin the ground temperature drops steeply, but the caribou can expose forage by scraping with their hooves. Where snow accumulates in drifts, lemming and ptarmigan can den in the warm zone above the frozen ground. Snowdrifts, slow to melt in summer, shorten the growing season but provide a constant source of water to the plants colonizing the ground beneath.

PERMAFROST

At high latitudes (and at high altitudes), where the mean annual air temperature is below freezing, the ground temperature in the near-surface portion of the earth's crust remains near or below the freezing point of water indefinitely. The ground, except for a thin surface layer thawed each summer by the around-the-clock sunlight, is perennially frozen to depths sometimes exceeding a thousand metres. Ground whose temperature remains consistently below the freezing point of water is defined as permafrost. The term *permafrost* describes the ground's thermal regime. It has nothing to do with the soil type or the amount of water or ice present in the ground.

Permafrost underlies fully a fifth of the world's land mass, including half of Canada's. Most of this is continuous permafrost located beyond the tree line. Within the tree line the permafrost may be discontinuous, islands of perennially frozen ground surrounded by ground that freezes to a shallow depth in winter only to thaw again in the spring. Discontinuous permafrost has a precarious existence: two hot summers in a row or a human-induced change to the insulating properties of the ground surface are sufficient to alter the dimensions of the island. This sensitivity complicates the work of foundation engineers. Below bodies of water too deep to freeze all the way to the bottom a bulb of unfrozen soil co-exists with the surrounding permafrost.

Continuous permafrost creates a vast tabletop just below the ground surface that is impervious to water flow. Water cannot percolate through frozen soil—it can either transpire, evaporate, stagnate on the *permafrost table*, or run down to the sea in summer if the tabletop is sloped in the right direction. This means that plant nutrients remain locked in the frozen ground. Although the permafrost table runs roughly parallel to the ground surface, its depth varies locally with soil type and exposure to sun, wind, and water. In silty soil on a slope shaded by the brow of a hill and covered by a snowdrift much of the summer the permafrost table will be high—a few centimetres below the surface. On a dry, gravelly ridge exposed to the sun and swept clear of snow the table will be low—several metres or more below the surface. The depth of the permafrost table can be estimated by interpreting the land forms and vegetation cover visible at the surface. More precise measurements are made by digging test pits or drilling test holes.

When water in soil freezes, it expands, increasing the volume of the soil/water mix. When ice in soil thaws the volume shrinks. The water may be contained in the space between the grains of soil or segregated within the soil as an ice lens or both. Sudden and cyclical changes in soil volume create unstable conditions for foundations.

EARTH

Northern Canada has three major physiographic regions. The Laurentian Shield dominates the east and centre; granite and gneiss outcrops lie in rolling topography scarred by the grinding of the last great ice sheets. Freshwater lakes abound. The rock is very old—some formations north of Yellowknife are thought to be 3.95 billion years old (about 85 percent of the earth's life span). Rock bordering the east arm of Great Slave Lake contains fossil evidence of cyanobacteria—primitive life forms—3.5 billion years old. By comparison the earth's crust beneath the oceans is less than 200 million years old. Most of the shield extends only a few hundred metres above sea level. Its eastern periphery, indented by fiords and glaciers (Ellesmere, Baffin, Labrador), rises to levels on the order of 2,000 metres. The second major region consists of flat-bedded plains and plateaux, mainly limestone, centred on the Mackenzie River Basin, and of flat coastal plains made of ice-rich silty deposits centred on the Northwest Passage. Folded mountains, a northwestward extension of the Rocky Mountain chain, form the third region. Generally less than 3,000 metres in height and geologically complex, these mountains contain the upthrust, crushed remnants of successive collisions between tectonic plates. Farther south in the chain, on the British Columbia-Alberta border, the Burgess Shale site contains the fossil remains of soft-bodied organisms 530 million years old.

In northern Canada middle magnitude earthquakes are quite common along the continental periphery—that is, in the western mountains, the northern archipelago, and the eastern seaboard. Seismically, the centre is almost quiet.

PATTERNED GROUND

In summer the barrens terrain is naked to the eye. The patterns worked on the ground by frost, permafrost, rain, wind, water, ice, isostatic rebound, earthquake, exfoliation, gravity, weak acid, and biota are as plain to see as the dark seas on a harvest moon. But land forms are thousand-year work, the result of a long-term process, not an overnight intervention. Any physical damage done to this landscape takes scores of years to heal.

In fine-grained, ice-rich soils ice-wedge polygons, *solifluction* terraces, *pingoes*, frost boils, sorted and non-sorted mud circles, *thermokarst topography*, and *drunken forest* seem to compete for space in the landscape. All originate to some degree in the existence of permafrost and the freeze-thaw cycle. Ice-wedge polygons start life as a network of shrinkage cracks in the soil surface caused by exposure for long periods to very low air temperatures. Meltwater seeping into the cracks refreezes and expands, widening the crack wedge-like over many seasons. When the cracks become wide enough for sunlight to enter, the ice there melts, causing the soil to slump. Solifluction is moist, fine-grained soil creeping downhill along the slippery permafrost table in waves or small terraces resembling contours on a topographic map. When a deep lake drains for any reason the bulb of unfrozen soil below it will be retaken by the surrounding permafrost, working from the periphery to the centre. The frost's encircling action compresses the water-rich soil in the centre so much that an eruption of water may occur at the surface. There the water refreezes in increments over many seasons and forms a cone, perhaps as high as 65 metres, known as a pingo.

Similarly, the *active layer* refreezes each year from the surface downward and, more slowly, from the permafrost table upward. This puts a squeeze on the last-to-freeze soil caught in the middle. Excess water may burst through the surface as a frost boil. In large patches of silt and gravel the annual effects of freeze-thaw tend to segregate stones from silt in a process that "sorts" or pushes gravel and stones to the boundary of circular patterns about a metre in diameter. Thermokarst pits or lakes tens or hundreds of metres wide result from the local degradation of permafrost. The permafrost, once exposed for any reason to sunlight and high air temperatures, thaws progressively each summer, deepening and widening a pit indefinitely. Without the cohesion provided by frost, the edges of the pit slump. The trees growing at such boundaries lean inward at crazy angles for a few seasons before falling: drunken forest.

On coarser soils ranging from silt-free sand to exposed bedrock, the effects of glaciation may still be evident: glacial tills, moraines, *eskers, drumlins, erratics, roches moutonnée*s, striations. The Belcher Islands in southeastern Hudson Bay, seen from the air or on a map, appear to have been clawed from the sea bottom.

Seal on ice floe at ebb tide. Alexandra Fiord, Ellesmere Island—July

HUMAN TRACES

To resupply the glacier team on the lower reaches of the Sverdrup Glacier, the icecap crew establishes a cache somewhere on the upper reaches. The two of us in the glacier team take turns looking for the trove through binoculars, the one determined to find the cache and the other looking for any small irregularity in the immense landscape of ice and rock, light and shade. The former fails to see his target but the latter spots an anomaly. The cache is too far away to make out, but the presence of an anomaly is plain as day. What humans do to the landscape always stands out. (It takes extra mental effort to devise something that fits into the landscape.) The glacier team sets a new course to reach the anomaly, which finally resolves itself into a small fuel drum with a lengthy ice drill leaning against it. A tin of honey rests on top of the drum. These right-angled silhouettes stand out because they have no parallel in this landscape. This is the principle of the inukshuk, the cairn of flat rocks made by the Inuit to appear human from a distance, to act as a beacon for other humans or as a caribou-drive boundary marker.

Artifacts older than humankind clearly outnumber manufactured objects in the high latitude landscape. The patina of civilization is extraordinarily thin. This is the reverse of the urban environment, where a mature tree seems out-of-place in a field of concrete, glass, and asphalt.

Measured against the vastness of high latitude terrain, pollution caused locally mounts slowly—few people produce little garbage. But pollution imported from midlatitudes is already significant and out of control—food chain contaminants like dioxin and mercury are widely reported—since pollutants do not respect political and sociological boundaries. In North America and Eurasia many watersheds that tilt toward the Arctic Ocean have their southern limits at midlatitudes. Atmospheric circulation brings material released at midlatitudes to polar regions within days.

AIR AND FIRE

AIR OCEAN

Human beings live at the bottom of an air ocean. Most live right on its floor. Some, Tibetan and Andean shepherds among others, live higher up, where stiff currents swirl around pointed reefs. Either way, humans need solid surfaces on which to walk and play. Unlike fish we cannot swim or hover in this sea. Air—60 times thinner than water—offers less resistance to human movement but also less support. Air is more mobile, more responsive to change. Its mobility allows the surplus energy received from the sun at low latitudes to be shifted to the energy-starved high latitudes. (Sea currents such as the Gulf Stream do similar transfers.) The high reflectivity (*albedo*) of snow and ice at the ground surface and low sun angles limit the amount of solar energy absorbed at high latitudes. But air currents keep the earth's energy budget in a balance tolerable to life—air temperatures at the equator never reach water's boiling point and air temperatures at the poles never sink out of sight in the direction of absolute zero. Through a chain of convection cells warm air rising at the equator migrates poleward as cold air descending at the poles makes its way to the equator (Figure 1-06).

This responsiveness in the air ocean—the atmosphere—makes its upper surface as chaotic and choppy as the surface of the sea. Unseen waves follow unseen troughs. But large banks of cloud announce the coming of a trough and blue sky signals a wave. And the strange behaviour of light reaching earth from the night sky serves as a reminder of the turbulence in the atmosphere. Discrete points of light, seen through a lens whose density and thickness changes incessantly, become fitful starlight.

The earth and the moon share the same region of interstellar space and orbit a similar distance from the sun. But without an atmosphere to attenuate and redistribute solar radiation the lunar surface goes cold as soon as the heat is turned off. Exposed to sunlight again it heats up instantly. Surface temperatures vary about 200 degrees Celsius. On earth the air and water oceans act as buffers between these extremes. The buffers cause lag time, a delay between the day of maximum radiation received and the day of maximum heat felt at the bottom of the air ocean. In the northern hemisphere summer temperatures peak about a month after the June 22 solstice; winter minima lag behind the December 22 solstice about a month as well. The lag time between the solstices and temperature change in soil and sea water exceeds a month because soil and sea exchange heat more slowly. At high latitudes this means that the snowbound earth and the sea ice, which look colder in winter than the air, are actually warmer. In summer, the reverse is true: the earth and the water remain colder than the air. The temperature of the sea varies no more than 10 degrees Celsius all year. The air temperature at ground level may vary 80 degrees Celsius. Buildings, which spend their working lives in the air ocean currents above the ground, must cope with the larger temperature range.

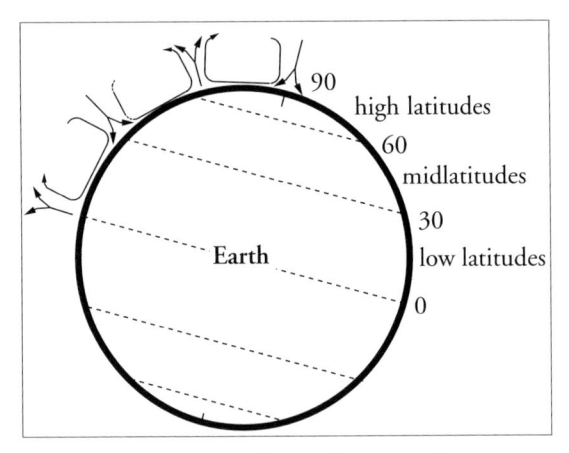

Figure 1-06 Air circulation between the equator and the poles.

SOUND

Sound travels far in cold, dense air. On a calm day at -40°C conversations may be overheard at distances up to a kilometre. Human voices at such distances appear disembodied. The barking of dogs can be detected at distances up to 15 kilometres. Animals preyed upon by humans overhear the hunter's approach over drifted snow from well beyond rifle range—a factor in many tales of death by starvation. Since optical effects often obscure vision at such distances, one may hear events long before seeing them. The hum of an approaching aircraft may precede sighting by ten minutes. But once a discrete event comes within earshot and eyesight an opposite effect becomes apparent. Since sound travels only 331 metres per second in dry air at sea level, or about one kilometre in three seconds, and sight is nearly instantaneous, the eye registers distant events before the ear. By the time a gunshot is heard at a distance of one kilometre the hunter has bent over his *komatik* to sheath his weapon.

On the barrens there are not many objects capable of attenuating sound. There are no birch stands, wheat fields, or city blocks to absorb airborne vibrations. The noise of a stream ranges far beyond the stream bed and the noise of a diesel generator can envelope an entire village. In a landscape where quiet has prevailed for eons small miscalculations in acoustics can cause noise pollution worthy of a big city.

AIR QUALITY

Away from the dust of mine shafts and the streets of settlements the air found at high latitudes contains hundreds of times less particulate matter than air breathed at midlatitudes. Polar air seems so clean, so fresh, that it could be used as a standard or background level against which to measure air pollution caused by industry in the rest of the world. The notion that air is cleaner at the poles than anywhere else first came under attack by pilots who, for most of the year, can see five times as far in the Arctic and Antarctic as they can in the tropics. Their 1950s encounters with high level haze in the arctic winter prompted scientific studies twenty years later. Weather stations in northern Canada and other polar countries now routinely monitor arctic haze. In Canada high-level haze has been reported as far south as 65 degrees north latitude. Industrial pollutants generated in northern Europe find their way to points over North America on air currents. (Midlatitude North American pollutants travel in the direction of Europe.) Summer winds scatter these pollutants, but in winter the polar air mass remains comparatively stable. The pollutants stagnate, much like smog over a big city. Arctic precipitation is too slight to scrub the air immediately. Haze components include synthetic chemicals such as polychlorinated biphenyls (PCBs) and pesticides, plus natural materials such as sea salt, soil, and pollen. They increase the acidity of rainfall, cause toxins to accumulate in the food chain, and screen solar radiation sufficiently to alter the polar energy balance.

Two decades of atmospheric research have shown that the density of high level ozone changes continuously. The ozone layer screens most of the ultraviolet radiation known to be harmful. The traces of ultraviolet light that do penetrate the screen cause sunburn, skin cancer, and eye damage in humans, and a reduction in plant growth on the land and in the sea. Steady degradation of the ozone layer would radically increase the dosage of ultraviolet light.

Coronation Gulf and village dump. Kugluktuk—July

OPTICAL PHENOMENA

At mid and low latitudes humans rely on atmospheric perspective to gauge the distance of faraway objects. Distant objects seem less distinct and darker than objects in the foreground—they are lost in a purple haze.

At high latitudes, the recent phenomenon of arctic haze notwithstanding, there is not enough moisture, dust, pollen, and smoke to provide atmospheric perspective. A distant mountain seems as clear and as detailed as a nearer one (partly because the human eye sees the large detail of the distant mountain as the equivalent in size and shape to the small detail of the nearer mountain). Without atmospheric perspective in the line of sight the distance between remote objects appears to collapse—background and foreground run together. One ridge of hills cannot be distinguished from the second and third, so a three kilometre ridge-to-ridge separation

Twenty kilometres to the horizon. Bache Peninsula, Ellesmere Island—July

appears to be zero. Outsiders frequently underestimate distances at high latitudes.

Light passing through one fluid (cold, dense air) into another (warm, thin air) changes direction slightly. The light refracts. At low and midlatitudes, air temperature and density usually decrease with increasing altitude. At high latitudes, where air masses may remain stable for many days at a time, temperature inversions occur frequently. The air closest to the earth's surface may be colder than the air a few hundred metres higher up. So light arriving from the horizon may pass through pockets of air with different densities—it refracts, distorting the image in the line of sight. At high latitudes the solar disk appears to remain above the horizon well after it has actually set. At low latitudes the disk disappears as though yanked. Refraction also means that a distant band of cliffs that should be out of sight below the horizon straggles into view above the horizon, shimmering, insubstantial, unconnected with earth. Land masses and icebergs made weightless in this way are called "*loomings.*"

A looming is a large mirage. Smaller mirages occur on bright, sunlit surfaces where air density and wind speed vary continuously. The shimmering image of a rampant polar bear seen far out on the sea ice turns out to be a sea gull on closer inspection. The familiar face of a steep river bank, seen at a distance, becomes an insurmountable ridge. Icecap surveyors curse the sunny days—their targets shimmer and scintillate inside the eyepiece, defying optical measurement—and relish the overcast days.

From the flight deck, six kilometres above Davis Strait, we marvel at the four suns ahead of us to the west. There must be a fifth below the nose of the aircraft. The main sun, the familiar one, occupies the centre of the pattern. Of the others in view, one is above, one is to the left, and one is to the right. A halo of huge circumference, as golden perhaps as angels, links the outboard suns. Neither pilots nor passengers have ever seen such an awesome display. All the while, ice crystals stream past the windscreen by the millions. We are seeing the sun—and sundogs—*through an enormous atmospheric lens made of ice crystals suspended in the atmosphere. The images recall rainbows, except there is no separation of colour. The bogus suns are as white hot as the original. At 140 metres per second we rush toward the centre of the pattern but make little progress. Thirty minutes later, crestfallen, we give up the chase and descend through a peasoup overcast to land in the angel-free gloom at Iqaluit.*

NORTHERN LIGHTS

The Latin term, *aurora* borealis, means northern dawn—aurora australis in the southern hemisphere. Photographed from a satellite in polar orbit, the aurora borealis appears as a band of light forming a huge oval centred on the geomagnetic north pole. (The geomagnetic north pole presently resides over northwest Greenland. The magnetic north pole, the pole that dips the compass rose, lies north of Bathurst Island for the time being.) In Canada the band, usually 500 to 1,000 kilometres wide, stretches from about 65 degrees north latitude in the west to 55 north latitude in the east. Occasionally auroral activity takes place even closer to the equator. Although the aurora occurs in the ionosphere 100 to 300 kilometres above the earth's surface, it is produced by the solar wind, a shower of atomic particles first impacting the earth's magnetosphere at an altitude of 5,000 to 10,000 kilometres. Arriving at high speed, some of the protons and electrons penetrate the earth's magnetosphere near the geomagnetic poles to excite the atoms and molecules of the rarefied upper atmosphere. The surplus energy is quickly released again. Visible only on the dark side of the earth, the energy release is signalled by light emissions. Atomic oxygen (one atom instead of the usual two) shows green or red light. Molecular nitrogen gives off purple.

VOLCANISM

From time to time the situation on the ground or in the sea obliges Icelandic cartographers to add or subtract territory from their topographic maps. Iceland's land mass, almost entirely volcanic in origin, lies over a major sub-Atlantic intersection of tectonic plates. The plates are separating here at the rate of two or three centimetres a year. The rule along the gap between the plates is gentle eruption of molten rock hundreds of metres below sea level; the exception is explosive. On November 14, 1963 an undersea explosion southwest of Iceland raised Surtsey, a cone of volcanic ash that for a few days competed with the waves for supremacy at the surface. Surtsey won when it produced new rock to replace the easily eroded ash. But other explosive vents nearby came and went with astonishing frequency for several months. Two years later Surtsey calmed down enough to start growing plants and nurturing small animals on its own. Then the cartographers moved in with echo-sounding machines to map its undersea contours.

There is no such excitement in Canada. There are no active volcanoes. Some ancient ones in the east (Montreal's Mount Royal, for one) have been ground down to nubs by the continental ice sheets. The few in the west remain quiet despite intermittent volcanic activity just across the border in Alaska and Washington State and Canada's membership in the Pacific "ring of fire," the chain of active volcanism girdling the Pacific Ocean. But just outside of Whitehorse, Takhini Hot Springs offers a swim in a geothermal pool, winter and summer, to prove that Canadian volcanism lives.

Lot separation and fire crew saved neighbouring buildings. Rankin Inlet—March

FIRE

In the lee of a spruce fire the sky fills with black smoke and a resinous stench that shrivels nasal membranes and makes lungs heave. The pungent, invasive smell of burning sap permeates clothing and frays patience. Closer to the heat major damage is being done: buildings and electrical grids are threatened or demolished, trap lines are wiped out, streams are contaminated, permafrost tables are reorganized. But in the long term, renewal of the ecosystem follows. Nutrients trapped in old wood are returned to the soil, fresh plant growth feeds more herbivores than before, smoke particles become nuclei for the formation of rain droplets, scorched timber makes dry firewood.

For a forest fire to start, hot summer air, dry combustibles, and ignition must combine. Lightning—strikes reaching the earth's surface number in the billions every year—often provides the ignition. The bases of storm clouds become negatively charged while the earth's surface below remains positively charged. A discharge from the cloud, travelling on charged particles of air, meets a twin travelling upward from the ground and explodes in a flash. Lightning strikes an estimated 100 people per day worldwide. It strikes many more trees, hills, and buildings. It readily ignites woody or peaty landscapes that have been dried out by the sun and wind. Yet half of North American forest fires are set by itinerant campers. Orbital remote-sensing machines have been invented to locate lighting strikes, but there is no progress in pinpointing careless fire-makers.

Forest and brush fires are common to the Canadian subarctic, the *boreal forest*. There is not enough dry combustible on the barrens to constitute a chronic fire hazard. Most forest fires are left to burn themselves out. Some continue to burn underground for weeks, turning the soil to ash, which is then lost to erosion. Others, closer to human settlement, are fought rigorously, providing seasonal employment to hundreds of citizens.

Fire in buildings at high latitudes, particularly in winter, holds the special terror that shelter lost within an hour may takes months or years to replace. When a cooking fire destroys the only tent on a three hundred kilometre excursion there may be no happy outcome. Fire extinction services in remote high latitude settlements, based on minimal equipment and the work of trained volunteers, can usually do little more than prevent the spread of fire to neighbouring structures. Physical separation—keeping a minimum distance between buildings clear of combustibles—remains the principal means of limiting loss by fire in high latitude communities.

7 FLORA AND FAUNA

DESCENT TO THE OASIS

The snow is too wet for a ski plane to land. The nearest helicopter loafs on an icebreaker's afterdeck 200 kilometres away, and is, we are told, too dainty to do menial labour at this altitude. Our camp lies at the 1,000 metre level on the northwest corner of the Devon Icecap. The summer's work complete, we are impatient to rejoin our colleagues at the base camp on a lowland by the sea 30 kilometres to the west. From there our chances of being flown back to southern Canada are better than even. From the icecap they are nil. We are ready to walk the first leg home. The way down the Sverdrup Glacier, too heavily crevassed in August for safe passage, ends at a point on the coast too far removed from the base camp. That leaves the direct route down to the edge of the icecap and over the plateau.

The edge of the icecap is like a bell curve, round at the top, steep on the side, and flared at the bottom. The wet snow underfoot makes travel down the curve laborious—we leave deep footprints in winter snow that failed to melt during this year's cooler than usual summer. Finally we step off the icecap onto the plateau, onto ground more solid than we have felt in weeks. Released from the snow our legs feel like coiled springs. It is also strange to be walking without leaving footprints. The plateau consists of frost-shattered bedrock—there are no plants, not even lichen to darken the buff-coloured rock. The stream beside us takes meltwater from below the icecap down to the sea. Freezing cold, the water is milky with silt. Hours farther down we run afoul of a web of gullies. We slide into one gully and come to rest on a dirty snow bank that sags as the thaw progresses. Then we scramble with little grace up the opposite side only to find another gully across our path. At last the creases in the terrain peter out and the plateau changes from rocky ground with patches of mud to muddy mesa with patches of broken rock. Footprints reappear in our wake.

We have not seen birds for weeks, nor insects. Here and there some tufts of greenery beat the odds of polar desert life at high elevations. But the few flower stems visible have long since wilted.

Even with hours of muddy plateau behind us the perspective ahead scarcely changes. To relieve the tedium we turn off the plateau into a rift valley. The sound of flowing water breaks the silence. Everything around us seems to be rock or water. But there are more signs of life. The green patches are larger. Having forgotten how lush greenery can be we stare and stare, becoming unsteady on our feet as precipice and scree *replace plateau. In an hour the south valley wall, sheer, vertical, uplifted, and black, races off to the west horizon, tracing the fault line and cradling a deep inlet of the sea. The north valley wall, the low, downshifted, buff-coloured one, gives way to the lowland plain beyond.*

What an Eden lies before us! Life here jostles for space. In the dusk below a stratum of metal-grey cloud, raised beaches, freshwater lakes, ponds, and sedge meadows lie stranded in a shallow sea of vegetation. Greenness of a striking intensity and variety assaults retinas until recently embalmed in the snow white, rock grey, and sky blue of the icecap. Greenness seen close becomes stalks of grass spearing tufts of wind-blown eiderdown. Ankle-height willows cling to beach gravel. White foam whipped up by the wind laps against the lee shores. Patches of reindeer moss—really a lichen—mix with patches of real moss, green, moist, and deep. Yellow poppy, purple saxifrage, orange-, lime-, and black-coloured lichen thrive among the scattered boulders and exposed traces of charcoal-grey bedrock. No footprints here. The rock is too hard and the moss too resilient to note the passage of feet. The lowland is roughly square. Two of its sides end by the sea in a froth of breakers and uprooted seaweed; a third abuts the inlet by the rift valley wall; and the fourth stops at the foot of the bleak plateau we vacated moments ago. All traces of snow on the land and ice in the lakes have disappeared. Ahead, by the last lake in a series, the rectilinear shapes of the base camp emerge from a backlit sea mist.

The midnight sun is finished for this year. Evening darkness subdues the landscape. Shortly twilight returns and every migratory bird on the tundra gives voice. An intoxicating cacophony reverberates around us as though to confirm that, indeed, paradise is right here. Snow geese dot the grassy wet meadows. Woolly muskoxen graze the backslopes of raised beaches. Fox and weasel bounce in and out of cover. Walruses waffle in the roiling water offshore. The white sails of icebergs compete with loomings at the horizon. Bleached white as icebergs, the scattered bones and skulls of muskoxen poke through lush carpets of moss. Nearer the seashore among rocky outcrops more things poke through patches of moss and grass: this time remnants of Thule house circles. The arctic oasis.

Muskoxen grazing beyond a raised beach at Truelove Lowland. Devon Island—August

THE TUNDRA

A Lapp word, tundra, denotes polar plain. Tundra is the dominant vegetation region of northern Canada, extending from the arctic archipelago in the north to the edge of the *taiga,* the boreal forest, in the south. Vegetation cover is sparse—next to zero toward the pole but nearly continuous at the southern limit— and meagre in total biomass. Many of the species present have relatives at midlatitudes. Compared to lower latitudes the number of species is small, limiting the number of possible relations between species. A simple, unstable ecosystem results rather than a complex and stable one. Decline in one species directly affects all the others. The ecological balance is easily upset and slow to recover, despite the burst of biological activity in the summer.

Consider the constraints to growth. Precipitation in the polar desert averages about 100 millimetres annually. Poor distribution due to snowdrifting and poor drainage due to the imperviousness of the permafrost table located 20 to 60 centimetres below the surface limit the benefits of scarce precipitation. The permafrost table prevents deep root penetration and keeps the shallow active layer cold enough to retard absorption of water by the roots. The heaving and settling caused by the freeze-thaw cycle keeps the fine-grained soil of the active layer in a state of turmoil. Low winter temperatures and the drying and abrading effects of wind contribute to high mortality rates. The nutrients in the soil essential to sustained growth, especially nitrogen and phosphorous, are in short supply. Low temperatures inhibit the decay of plant and animal detritus, resulting in long nutrient cycles and high soil acidity. Low temperatures also prevent polar flora from contributing oxygen to the atmosphere and absorbing carbon dioxide in significant amounts. The mean air temperature of the warmest month is less than +11°C. The growing season of one to three months is short and the angle of sunlight is low. Shadows are long. The path of the midnight sun, however, does expose plants to sunlight from many points of the compass.

Plants in the polar regions seem to select habitat with meticulous opportunism. They thrive in their favoured niche but falter in an adjacent one that has slightly different soil characteristics and exposure. Crustose lichens, saxifrage, and the occasional arctic poppy cling to dry beach ridges. Arctic willow, grasses, heathers, and mats of lichen and moss occupy the *tussocky* backslopes of those ridges. Sedges, herbs, and other mosses proliferate in the wet lowland between them. This sensitivity to microhabitat becomes a vivid spectacle for a few days in the fall when the leaves turn. The tundra flashes red, orange, burgundy, and rust in brilliant profusion, showing, with elegant precision, a discrete change of hue just where the habitat boundaries lie. No Persian carpet could be more intricate or more beautiful.

Specific adaptations improve conditions in the microhabitat. Most polar plants hug the ground, their stems and leaves forming a canopy that deflects the wind and absorbs sunlight efficiently. Temperatures in the cosy world below the canopy are several degrees higher than in the air currents above. Dryas flowers, shaped to concentrate sunlight, turn to follow the path of the sun around the horizon, thereby raising the temperature of its critical organs by several degrees Celsius. Some plants grow hair as insulation. Others have frost-resistant structures. Still others locate below perennial snowdrifts to benefit longest from meltwater supply. All such plants are accustomed to temperature variations up to 50 degrees Celsius. In contrast, many tropical plants wither when the variation exceeds five or ten degrees.

The tree line, the fabled natural boundary of polar and alpine regions, is less distinct than a political boundary but nearly as mobile. It is a transition zone kilometres or tens of kilometres wide where tundra vegetation, mostly reindeer moss, runs into the taiga like a high tide. The taiga's dwarf spruce or pine pokes up through the flood like clumps of mangrove. Climate changes measured in hundreds of years cause the tree line to shift poleward or equatorward. A thousand years ago, for instance, the tree line in Canada existed a hundred kilometres farther north. The frozen, not petrified, remains of giant redwood trees thirty-five million years old found recently on Axel Heiberg Island suggest climatic variation so radical that the twitching of today's tree line pales by comparison.

The tundra is vital to the food chain leading to humans. It feeds caribou. It feeds birds. It nurtures insects, food for fish and birds. It provides breeding habitat for waterfowl and lemmings. It provides plants for medicinal use. It provides Vitamin C in berries and saxifrage flowers to itinerant pickers. It provides lichen and dwarf willow for countless camp fires. Refuges and national parks, by protecting wildlife breeding grounds, safeguard a small proportion of the tundra. But industrial society tends to view the tundra as a backdrop for tourism sites, a platform for the extraction of mineral resources, and a reservoir for pollution. Settlements and townships in Canada develop without regard for the land. They obliterate the tundra.

THE TAIGA

The taiga (Russian for boreal or northern forest) has joined the endangered list. Extending from the tree line well into midlatitudes it is the largest vegetation region in Canada. In southern Canada it is harvested for lumber and pulp. Hardy conifers dominate the landscape—black spruce, pine, larch, fir. Stands of birches and alders occupy patches of well-drained ground. Mature tree size varies greatly with microclimate. The undergrowth is dominated by mats of fallen needles, by lichens and moss in profusion, or by berry-producing shrubs.

Discontinuous permafrost underlies much of the taiga. Permafrost, the long winter, and needle litter from the confers retard soil development; the active layer remains thin, acidic, and poor in nutrients. The return of nutrients to the soil through decay is also slowed. In areas of especially poor drainage, layers of partially decayed plant matter are deposited over previous layers to form peat bogs whose cover is limited to black spruce, evergreen shrubs, and sphagnum moss. Disturbances to the peat surface (such as uprooting of a tree) result quickly in exposure of permafrost to warm air and sunlight. The ground slumps and pits form in the resulting thaw.

Precipitation in the taiga is two or three times that in the tundra. In winter, the forest deflects blizzards over the tree tops so that snow drops out of the airstream and settles in thigh-deep, fluffy accumulations among the trees. The trees shade the snow from sunlight as well as wind. It remains there with little external change until the spring thaw melts it in a rush, flooding streams and rivers. In summer the long solar day increases the intensity of the short growing season—Yellowknife spring starts only at the beginning of May, compared to early April in Montreal, but by the end of May, foliage development in both places may be the same. Later the trees dry out—their dark colours absorb the solar heat.

Orchid on the right bank below Virginia Falls. Nahanni River—July

The wind still flies over the top, the forest now providing sheltered airways for mosquitoes and black flies. Dry, resinous wood makes fine tinder for forest fire. Fire accelerates the return of nutrients to the soil necessary to new growth but plant succession is slow, initially favouring birch and aspen. Fall rains and cold temperatures turn the leaves of deciduous trees yellow, a colour made vivid by the dark green backdrop of the conifers. Later, when winter fog coalesces on tree branches, the dark green turns a sparkling white.

In northern Canada products of the taiga have had traditional uses in medicine, food, and shelter. Trails through the forest, kept open by generations of hunters and trappers, still provide access to food and fur for families living close to the land. A limited quantity of timber is harvested for lumber products. Trees are also cut, less frequently now, to make log walls for houses, community buildings, smoke houses, and bare poles for tipis, tents, and storage platforms. In places that have been prospected for oil and gas rectilinear cuts crisscross the taiga to mark the passage of seismic survey crews. Similar cuts for winter roads to remote villages and mine sites link en route lakes end to end.

THE OWL AND THE RAVEN

In good hunting weather the snowy owl perches on a rock standing just a few tens of centimetres above the tundra surface. From there it can survey movement on all sides without projecting its own silhouette too far. There, body rigid, head swivelling, great yellow beacons for eyes, it scans the foreground for prey and the background for danger. Brown plumage mottles the snow white of its body. Bright orange lichen covers the rock favoured as perch and a mat of lush green moss surrounds it. This tiny oasis subsists on a rain of nutrients released by the decay of countless owl droppings.

After a long day of mapping vegetation I amble back to base camp, eyes glued to the tundra microworld unfolding underfoot. From time to time my eyes veer up to judge the way ahead. Half way back, something on the horizon catches my attention. Clearly not hunting and nowhere near a typical perch, a snowy owl stands on a patch of low ground, its back to a boulder two metres in height. Screaming at the top of their voices two birds dive repeatedly at the owl, in a cycle of near hits that make the larger bird weave and duck like a stricken boxer. The owl seems to be using the tall rock to block aerial attack from the rear. Its attention totally absorbed by the birds hovering overhead, it has missed my approach. Turning quickly to the right I run silently over the tundra, keeping the rock between the owl and myself.

At a range of a hundred metres the attackers turn out to be a pair of long-tailed jaegers. From their vantage point they make note of me and at fifty metres, judging my presence too hot to handle, they break off the attack and depart. Although blind-sided by the rock, the owl must wonder what would cause the jaegers to flee. Indeed, the abrupt silence shocks the senses. Now immobile behind the rock I imagine hearing the owl consider its next move while, on another channel, I listen aghast to the banging of the valves in my heart.

There is no further movement. To break the impasse I burst out of cover with a shout, flapping my arms like a snow goose working for lift. Even with its back turned the owl's eyes lock on me as I emerge. The huge bird leaps into the air, away from me, bending its wings to make distance rather than height. Three metres out its head swivels back half circle, wing motion uninterrupted, to judge the speed of my takeoff. At ten metres the head turns back again. And once more at thirty. Convinced at last that I am either flightless or feckless, the owl settles into ponderous flight low over the tundra.

The snowy owl avoids human settlement. The raven does not. Ravens gambol on the updrafts near the tops of highrise buildings. They leave droppings, white most of the year but purple in berry season, on the facades of any human edifice. Ravens have an eye for things stringy and elastic. They remove flexible foam rod from the bottom of construction joints. They toy with cured strips of silicone sealant found in roof construction, until a piece breaks off and can be flown to a new location. Any roof consisting of a thin plastic coating over polyurethane foam insulation will be pecked in aid of beak and neck muscle development. Ravens poke holes in satellite receiving dishes and in the skin of small aircraft. With a wing span over a metre ravens may short circuit high tension lines. They are a collision risk at airports. They make tourists turn and stare. Ravens are denizens of garbage dumps where refuse and rodents provide ready meals to the omnivorous. They are adept at opening bags and boxes to scavenge contents. They may make villagers crazy enough to vote for hiring a raven catcher.

Ravens are canny hunters. One bird of a pair will fly ahead to flush out a nesting shorebird. The second follows at a distance to spot the point of takeoff and plunder the camouflaged nest. Ravens will follow a wolf pack to pick at the remains of carcasses. They are long-lived and may mate for life. Ravens nest in rocky cliffs. Their young grow to nearly full size in six weeks. Their plumage is a lustrous black, like raven-coloured hair. They have two thermostats, one for the body proper and one for the legs, so that circulation temperature in the latter can drop almost to freezing without loss of function. The raven's omnipresence in the Canadian north, its opportunism, its longevity, its intelligence, and its raucous speech—ravens used to speak Inuktitut until condemned to speak otherwise for committing an act of greed—make it a major element in native mythologies going back to creation. (Today some researchers believe that raven speech varies across the circumpolar north—that ravens have regional dialects.) In Dene legend the raven was the first bird to appear at

Creation. Although ravens could not be trusted—they were tricksters—they sometimes predicted disaster. And sometimes they would lead starving hunters to game. Being commonplace and unfathomable at the same time, their purposeful flight and haunting voice readily penetrate the recesses of human minds.

OTHER ANIMALS

To industrial society undomesticated animals are "wildlife." To aboriginal society animals are not wild at all, animals and humans are inseparable parts of a single world. This fundamental difference in world view permeates and churns transitional societies at high latitudes. A comparison of the possessions found in mid- and high latitude back yards starkly illustrates the difference. At midlatitudes: garden tools, lawn mower, barbecue grill, pesticides, mouse traps, bird feeders, golf clubs, tennis rackets, skis. At high latitudes: snowmobile, snowmobile parts, sled, fuel containers, rifles, frozen caribou parts, bear skins, fish drying racks, skin stretchers, frozen whole seal, fox traps, rope, fish net, skates. The difference may be dissolving, slowly. Industrial society's recent call for the intact preservation of wildlife and wildlife habitat has been superseded by the more pragmatic view that wildlife must be managed scientifically as a renewable resource for the benefit of all. This view arrives with little time to spare, given that pollution trends and human mobility have virtually forced intact habitat and undisturbed wildlife out of existence.

Arctic denizens have adapted to local climate and geography in a complex variety of ways. Some control heat loss in winter by curling up into a ball, reducing surface-to-volume ratio. Hollow guard hairs in the fur of others enhance resistance to heat loss. Caribou ready to calve select ground exposed to sunlight, and to the prevailing wind, which keeps the mosquito hordes at bay. The fur of arctic foxes, ermines, and lemmings changes to white in winter, camouflaging their silhouettes against the snow. Some birds regulate body temperature downward. Some birds store food in their crops for consumption at night. In winter whiskey-jacks eat food they have stored in summer. Wolves and foxes den in the well-drained, ice-free soil of eskers left by the continental ice sheet. In summer beavers eat new wood in adjacent lakes so that the new wood in the home lake can be reached easily during winter. Some observers believe the beaver to be the greatest race of builders after humans, and certainly the oldest.

Bison, moose, caribou, seal, hare, and fish have been the mainstays of aboriginal existence for generations. It is still the case. Many families in northern Canada continue to harvest these animals for food, since packaged food, bought in general stores, remains high in price and low in nutrition. Muskrat and beaver fur has been central to the Dene subsistence economy for at least two centuries.

Human settlements frighten most wildlife: the ringing of church bells is enough to disturb seals. Vehicular traffic on the sea ice at the heads of some Greenlandic fiords is banned to minimize disturbance of seals. (Ironically, taped music played loudly in the bottom of a boat will pique the curiosity of nearby seals, making them easy prey for rifles.) The snowy owl and the gyrfalcon keep their distance. But the line between the new settlements and surrounding wildlife is blurred. Caribou are known to huddle in the warmth released by oil pipelines. Arctic foxes may adopt snowmobile tracks as highways. Birds used to nesting in trees may nest in building crevices beyond the tree line. Flightless baby gulls have been toys to Inuit children for centuries. Settlements, through exotic smells and garbage dumps, also tend to attract the kind of wildlife that cannot, for one reason or another, be counted on for food: bear, wolf, fox, sea gull. The larger animals pose a danger to the unwary. Unfortunately, the sedentary lifestyle of the village makes a growing number of people unaware of events beyond the perimeter.

Summer harbour. Illulissat—August

III
CLIMATE

8 SUNLIGHT
TRADITION *41*
PROBLEM *41*
CONCEPTS *41*
NIGHT AND DAY *43*
SUNSHINE *44*

9 TEMPERATURE
TRADITION *45*
PROBLEM *45*
CONCEPTS *45*
AIR AND SOIL TEMPERATURE *47*
DEGREE-DAYS BELOW 18° CELSIUS *47*

10 WIND, RAIN, AND SNOW
TRADITION *48*
PROBLEM *48*
CONCEPTS *49*
MEAN WIND SPEED *51*
WIND CHILL *51*
TOTAL PRECIPITATION *52*
MEAN SNOWFALL *52*
BLOWING SNOW *52*
SNOWDRIFTING *52*

11 HUMIDITY
TRADITION *53*
PROBLEM *53*
CONCEPTS *54*
SOURCES *56*
EFFECTS *57*

40 CLIMATE

*Disko Bay.
Qeqertarsuaq—August*

Rankin Inlet—March

8 SUNLIGHT

TRADITION

More intense every day, spring sunlight glances off the snow and embeds itself in the retina. There is no escape, the tundra and the sea ice offer little shade. Prolonged exposure leads to snow blindness and excruciating pain, the only antidote being a damp blindfold and three days of rest. In the spring, snow goggles, two horizontal slits in a strip of ivory or wood, are an essential piece of trail gear. In summer, because there is no snow cover to reflect the light, and in winter, because the sun is too low in the sky to matter, the danger recedes.

At high latitudes in the northern hemisphere the back gate of summer slams shut about the middle of August. One day the sun has a warming effect on the skin; the next, nothing: the sun is a wax image, powerless, girdled by a bundle of symbolic rays. Sunlight without warmth signals the profound change about to overtake living things in the polar landscape. Months later a pair of morning stars reappears to herald the sun's return and a fresh burst of light and life. In Ilulissat, West Greenland, 69 degrees north latitude, at noon every January 13, townspeople walk up the ridge overlooking the southern horizon to celebrate the reappearance of the sun.

PROBLEM

In northern Canada most buildings and communities are laid out as if the sun did not exist. (In some European countries access to sunlight is a legal right.) How can something so basic be overlooked, especially at latitudes where sunlight is scarce? Ignoring sunlight seems to originate in industrial society—a civilization that has declared its independence from the landscape—in part because fossil fuels afford reliable heat, even on cloudy days, rendering the sun obsolescent as a heat source for human habitation.

Ask children living at high latitudes why they play outside longer on sunny days. Direct sunlight moderates microclimates, warming air and surface temperatures well above the prevailing norm. Wind-sheltered space that traps sunlight becomes a fleeting patch of paradise. Space between buildings that ignores sunlight never comes alive, remains forgotten, and becomes a refuse depository, a symbol of shared neglect. Sunlight throws long shadows from buildings, blighting the space behind, and raising the profile of the permafrost table below.

Low-angle sunlight penetrates the depths of buildings having windows in appropriate locations. Sunlight is welcome in living rooms, shunned in sleeping spaces. It causes glare problems. The surface temperature of building materials exposed to sunlight can range from a low of -40°C in winter to +60°C or more in summer. Sunlight alters the chemistry of organic materials, reducing durability. On the other hand, solar radiation dries out wall and *cathedral ceiling* assemblies that have been wetted by rain or condensation, thus prolonging the life of the building.

Designers of buildings for high latitudes often ignore the solar heat gained through windows. The picture window demanded by midlatitude suburban tastes is a dubious feature at high latitudes. In the spring, sunlight streams through the glass, turning the room into a midday desert. The room, being super-insulated and built of light-weight materials, has no means of storing the excess heat for release later, when the sunlight disappears. At night, and in winter darkness, the window glass empties the room of warmth by conducting heat to the outside.

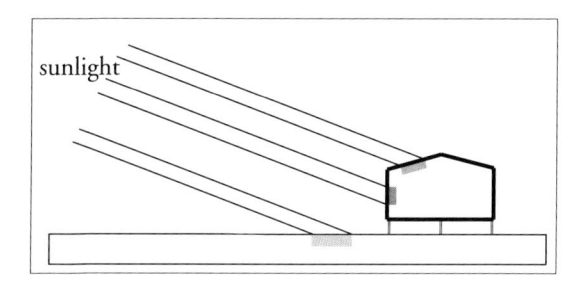

Figure 1-08 The concentration of sunlight received at a surface varies with the angle of incidence. Low-angle sunlight reaching a near-vertical surface (high angle of incidence) delivers more solar energy per unit surface area than does the same sunlight reaching a near-horizon surface (low angle of incidence).

CONCEPTS

The amount of sunlight reaching the earth's vicinity varies little, but the distribution of sunlight over a spherical object, the earth, varies considerably with angle of incidence (Figure 1-08). In a midlatitude summer the sun is high enough above the horizon each day that its main impact occurs on horizontal

42 CLIMATE

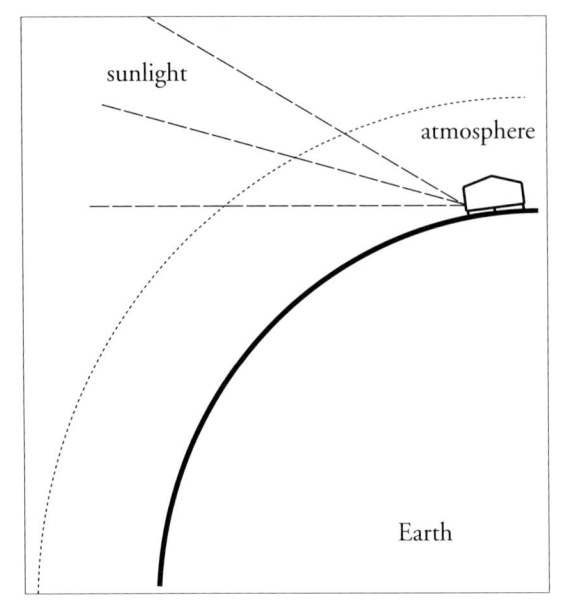

Figure 2-08 To reach the surface of the earth low-angle sunlight travels a longer distance through the energy-absorbing atmosphere than high-angle sunlight.

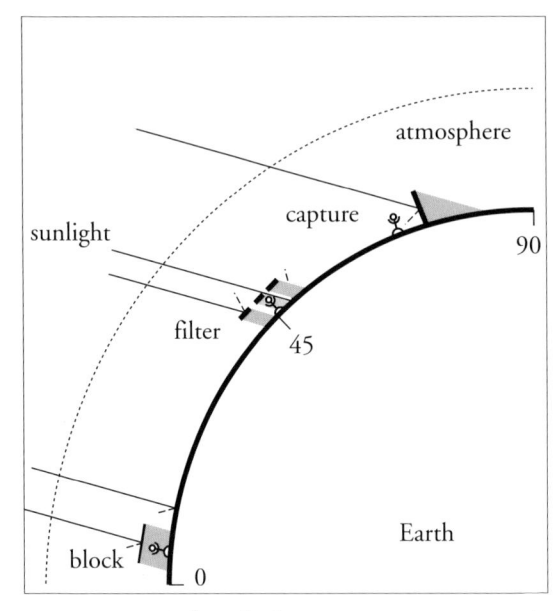

Figure 3-08 Solar shading strategies vary with latitude.

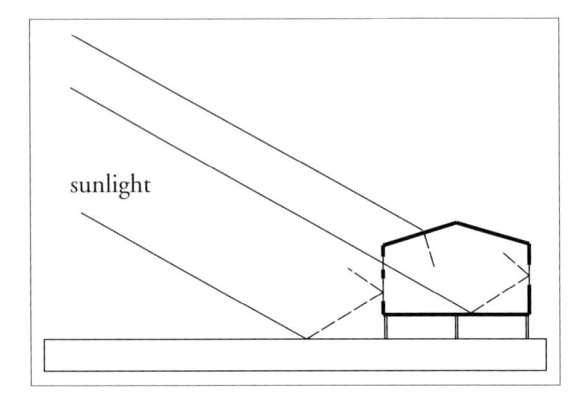

Figure 4-08 Being nearly transparent to direct sunlight but nearly opaque to the heat energy in indirect sunlight—heat emitted from a surface—window glass in buildings acts as a heat trap, a greenhouse.

surfaces—flat roofs, parking lots, and beaches. At high latitudes the angle of incoming sunlight remains low, so low that the sun's main impact occurs on vertical surfaces—walls and windows. But the low angle means that sunlight has to travel a longer distance through the energy-absorbing atmosphere. The thicker and the dirtier the intervening atmosphere, the weaker the sunlight that reaches the ground. For this reason the sunlight is most intense on walls and windows when solar altitude exceeds 25 degrees: its path through the atmosphere is shorter (Figure 2-08).

Solar shading strategies for building exteriors must vary with latitude. Figure 3-08 shows that at low latitudes, where the sun's path through the atmosphere to the earth's surface is shortest, complete shade, blocking out the sunlight, is the paramount concern all year round. At midlatitudes, where the direction and amount of solar radiation varies seasonally, the object is to filter sunlight when it is most intense, reducing its impact on materials and on cooling loads. Ideally the architect designs the filtering mechanism to operate fully in summer without preventing the penetration of sunlight in winter. (Deciduous trees placed on the sunny side of a building provide shade in summer and, leafless, allow sunlight to flood the building in winter.) At high latitudes, where sunlight is weak or non-existent much of the year, it must be captured and focused by vertical surfaces to provide a thermal comfort zone for people working or playing near a building.

Snow cover on the ground intensifies the light and heat energy reaching vertical surfaces; by midsummer, with snow gone, the effect subsides. This springtime energy bonus transfers little heat through walls that have been well insulated against heat loss during the sunless winter (Figure 4-08). Windows are different, since window glass is opaque to far infra-red radiation. They transfer much of the direct radiation to the interior, where it is absorbed and re-emitted at the longer wavelengths of infra-red radiation, radiation that cannot pass through the window glass: the "greenhouse" effect. Building designers must decide when and where a greenhouse environment is appropriate.

At high latitudes in summer the solar disk visits all or most of the horizon's circumference once a day. Sunlight penetrates the building or community layout from just about any point on the horizon. Detached buildings behave like 24-hour sundials.

Aklavik—July

Little Cornwallis Island—July

Although the sun's angle of incidence on flat roofs is slight, it is the flat roof that plays to the sun the most hours each day. Wall exposures and sloped roofs receive sunlight in turns as the earth rotates.

The ozone layer in the upper atmosphere screens most of the incoming ultraviolet light that is interactive with human skin and some building materials. The amount that penetrates does damage through prolonged exposure. Ultraviolet light alters the chemistry—the molecular structure—of materials such as plastics and rubber by leaching plastifiers or forming undesirable reticulation. The essential properties for which the material was selected may simply disappear.

At the present level of affordable technology, solar energy use at high latitudes is impractical for several reasons: sunlight intensity drops to nothing when heat is most needed; sun angles remain low much of the time; building materials commonly used at high latitudes, wood (low *specific heat*) and steel (low total mass), do not store heat energy efficiently; heat storage below ground conflicts with the need to stabilize local permafrost; there are no qualified maintainers.

NIGHT AND DAY

The 23.5 degree tilt of the earth's rotation axis with respect to the plane of earth's orbit around the sun is responsible for seasonal variation in climate. At one point along the earth's orbit the south polar region's exposure to sunlight peaks, leaving the north polar regions in the dark. Six months later, when the earth is half way around its orbit of the sun, it is the turn of the north polar regions to revel in the sunlight. The tilt of the earth's spin axis also produces variation in length of night and day. Between the Arctic and Antarctic Circles (latitude 66°32' North to latitude 66°32' South) day follows night every 24 hours, the day being longer in summer and shorter in winter. Between the circles and the poles, the length of the day varies from 24 hours to six months with increasing latitude—at the poles a six month day follows a six month night.

At low latitudes the length of each day hardly strays all year from just over twelve hours. Figure 5-08 shows daylight hours per day at midmonth averaged for several high latitude communities in Canada. (See also Table 2-31 in the Appendix). All points on the earth's surface receive about the same total amount of daylight each year.

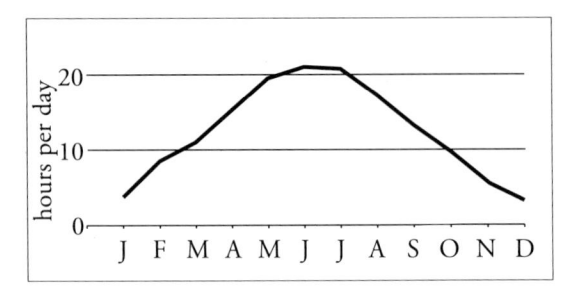

Figure 5-08 Daylight hours per day at mid-month averaged for eight northern communities in Canada.

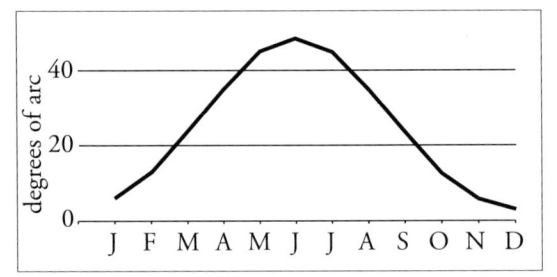

Figure 6-08 Solar altitude at noon, averaged for eight northern communities in Canada.

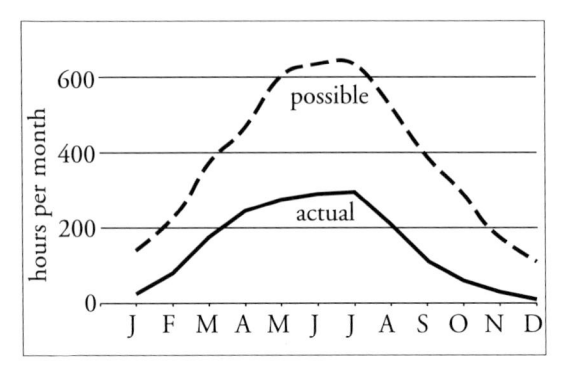

Figure 7-08 Possible sunshine hours per month versus actual bright sunshine hours per month, averaged for eight northern communities in Canada.

The altitude of the sun relative to a point on the earth's surface at a given hour on a given day (the sun angle) is calculated in degrees of arc, from zero at the horizon to 90 at the zenith. Solar altitude at noon decreases as latitude increases. Figure 6-08 shows the sun's altitude at noon during the third week of the month, averaged for the same high latitude communities. The summer peak at high latitudes is not impressive. In the tropics sunlight is nearly vertical. In polar regions sunlight is nearly horizontal.

SUNSHINE
In Figure 7-08 the total hours of sunshine possible, given the geometry of sun and earth, are compared to the actual bright sunshine received by several high latitude communities in Canada (Tables 3-31 and 4-31 in the Appendix). The maximum possible occurs at the Circles, not at the poles—Old Crow, Fort McPherson, Repulse Bay, and Broughton Island are the Canadian communities closest to the Arctic Circle.

The comparison between actual and possible sunshine is disconcerting. Of the potential sunshine high latitude communities in Canada receive only 40 percent of their due; cloud, fog, and blowing snow scatter the other 60 percent. Note that the two curves rise together in late winter and spring. Rising air temperatures and the absence of open water combine to minimize the production of cloud and fog. They are farthest apart in summer and fall, when free water at the earth's surface evaporates to obscure the sun with fog and cloud.

It is this effect, resplendent spring and cloudy autumn, that marks the turning of the seasons as much as summer light and winter darkness.

9 TEMPERATURE

TRADITION

Low air temperatures force people to conserve body heat, to take seriously the business of keeping warm. Each person must tend his or her own hearth to keep the flame inside alive. If the flame sputters there may be no warming sun, no wood fire, no hot food to boost body temperature again. Although the protection of a tent or snow house helps, only body metabolism and a space suit made of animal fur working together can keep body temperature near +35.6°C. A short, stocky body favours conservation of heat. A tall, thin one, with greater skin area for the same volume of flesh, does not. For those born to cool summers and cold winters, with tent and snow house for home, air temperatures near freezing do not provoke feelings of discomfort. The body acclimatizes itself. A sudden warm wind is more irritating than welcome. A person working comfortably outside for days at a stretch in air temperatures near freezing may be discomfited by a sudden rise in temperature of ten degrees Celsius. The space suit overheats; the walking surface turns to mush; mosquitoes, absent at temperatures below +3°C, mobilize quickly to look for blood meals.

So low air temperatures do have uses; others include cold storage of food and drinking water, and suppression of odour and rot.

PROBLEM

Some observers claim that high latitude cold is exaggerated, that Winnipeg can be just as cold as Baker Lake. This is true, but Winnipeg air does not remain at such temperatures for long. Baker Lake endures them for weeks at a time. Baker Lake people and buildings must adjust to continuous, not passing, deep freeze.

Low outside air temperature affects buildings in four main ways: heat loss, phase changes of water (gas to liquid to solid) in concealed spaces, freeze-thaw cycles at exposed surfaces, and dimensional changes of exposed materials (contraction). Two features of heat loss must be understood. First: the greater the difference between the inside and outside temperatures, the greater the rate of heat loss. The steeper the gradient, the faster the fall. Here Winnipeg on a bitter winter day matches Baker Lake on a typical winter day. Second: the total heat loss depends on its duration. If the bath is left running indefinitely the overflow will be great; in Baker Lake it runs for ten months of the year.

Low air temperatures condense the water vapour component of air into liquid and solid phases, either of which, when confined in stagnant air spaces, speeds up the deterioration of *building envelopes.* Water and ice in walls reduce the thermal resistance of insulations, and may induce wood rot when temperatures rise again. Ponded water and ice attack roof membranes mechanically through freeze-thaw contraction and expansion cycles. Foundation zones with high water content heave when frozen. Water pipes freeze; the fluid in fuel oil and propane gas distribution lines becomes sluggish.

Air temperature changes alter the dimensions of materials. Shrinkage cracks occur in many materials subjected to cold air temperatures. When chilled, the sheet metal facing on prefabricated wall panels may contract relative to its backing, only to buckle under expansion stresses when the temperature rises again.

CONCEPTS

Thermometers, thin glass tubes filled with a red bead of alcohol, are the simplest of analogue computers. They transmit a message—current air temperature—twenty-four hours a day, without supervision. We call the absence of heat "cold"; absolute zero—no heat—is about -273°C, or 273 degrees below the freezing temperature of water. The lowest air temperatures at the earth's surface do not drop below -80°C; that is, they remain about 200 degrees above absolute zero.

In the northern hemisphere cold polar air flows southeastward across continents, frequently bypassing continental west coasts. Humid air masses from the big oceans drift over west coasts, warming them. When the breadth of a continent separates a community from the modifying effect of the ocean to the west, the polar air remains unchallenged. In North America mean air temperatures drop not only as one travels north, but also as one travels east, away from the Pacific Ocean. The icecaps, the vestiges of the last Ice Age, reside in Greenland and the eastern arctic archipelago, not in Yukon Territory. There are exceptions to the rule. The ice fields in the mountains of the Yukon owe their existence to the cold air prevailing at high elevations, and to moist air originating over the Pacific. The Atlantic Ocean moderates east-

Arctic oasis. Alexandra Fiord, Ellesmere Island—July

ern seaboard climates. Where the polar air mass stands still for weeks at a time over the Yukon, cold air settles into the deep valleys as temperature inversions. (Usually the air at ground level is warmer than the air at higher elevations.) Snag, Yukon Territory, holds the North American record for lowest air temperature, -62.3°C. The January mean daily low at Snag is -35.6°C.

Low mean air temperatures maintain the stability of the permafrost regime in the ground. Any global warming would destabilize the permafrost, cause existing foundations dependent on frozen ground to settle, and release into the atmosphere large quantities of greenhouse gases such as methane presently locked in the ground.

In the northern hemisphere, air temperature response to seasonal variation in solar radiation lags about a month behind the solar solstice—January temperatures are lower than at the December solstice; July's are higher than at the June solstice. The atmosphere cushions the impact of sunlight at the earth's surface. This in turn delays the freeze-up and break-up of the waterways still more.

At high latitudes seasonal and daily shifts in air temperature vary too much for human comfort (from -50° to +30°C, a change of 80 degrees Celsius; surface temperatures of exposed building materials may vary even more). Designers conceive building enclosures (and space suits) to reduce the degree of variation to almost nothing, to stabilize inside air temperature within a preselected range of comfort (from +15° to +25°C, a change of 10 degrees Celsius).

Perennially low air temperatures slow the rate of chemical reaction between substances, corrosion, for example. For every 10 degree Celsius drop in temperature the rate of chemical reaction slows by one half. The rate at -40°C, for instance, is only 1/64th the rate at +20°C. This explains the slow deterioration of garbage at high latitudes. Wood rot at temperatures below +5°C is negligible.

Perennially low air temperatures at high latitudes reduce the frequency of freeze-thaw cycles compared to midlatitude winters. Nevertheless, freeze-thaw with water present is a principal weathering agent at high latitudes. In addition, building surfaces just below freezing temperature may be raised to thaw temperature by temporary exposure to sunlight combined with heat lost from the heated building. As soon as the sunlight disappears the temperature at the building surface drops below freezing again.

Since practical, cheap building envelopes cannot adjust chameleon-like to every shift in outside temperature, designers select a reasonable estimate of the severest outside condition for a given site as the basis for deciding the thermal resistance in walls and the size of the heating system. The National Building Code of Canada provides the 2.5 percent January design temperature for representative Canadian communities and towns. Only 2.5 percent of the hourly low temperatures recorded at a given location are lower than this design temperature. The January 2.5 percent design temperature for Snag is -51°C; for Baker Lake it is -45°C (See the Design Data table in the Appendix for a larger sample).

Compared to many other mammals humans are poorly insulated. Like other mammals humans give up heat through direct exchange of gases, liquids, and solids with their environment; humans also lose heat via infra-red radiation, and by conduction through contact with colder substances. Humans make an especially big splash on heat-sensitive photographic film; in contrast, the well-insulated

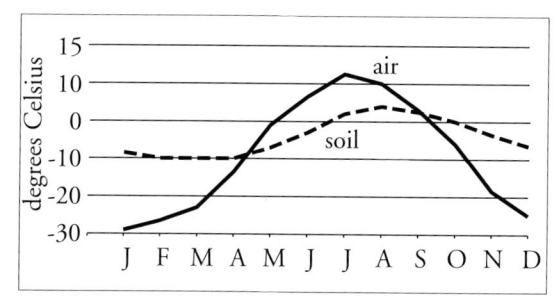

Figure 1-09 Mean daily air temperature versus soil temperature (one metre below surface), averaged for eight northern communities in Canada.

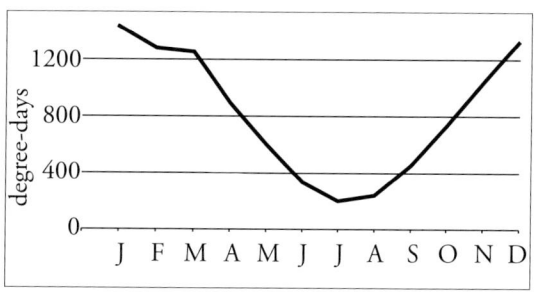

Figure 2-09 Degree-days below 18 degrees Celsius, averaged for eight northern communities in Canada.

Meltwater stream on the lower reaches of the Sverdrup Glacier. Devon Island—July

muskox remains nearly invisible. The ejection of water vapour from the lungs and the evaporation of water exiting the pores of the skin constitute the greatest loss. Uncontrolled heat loss leads to hypothermia and frostbite. But a temporary drop in body temperature of 3°C can be tolerated. Uncontrolled heat conservation leads to headache, nausea, and heat stroke. A rise in body temperature of 3°C can be life-threatening.

AIR AND SOIL TEMPERATURE

The lowest mean annual air temperature in the Americas occurs on the Greenland Icecap (-33°C); the air temperature at the North Pole, located over comparatively warm sea water, averages about -23°C over the year. Mean air temperatures in settled areas are considerably higher: -17°C at Resolute, -5°C at Yellowknife, -10°C at Whitehorse, and +10°C at Vancouver (Table 5-31 in the Appendix).

Mean soil temperatures (one metre below ground) are several degrees Celsius higher than the respective mean air temperatures. Figure 1-09 compares mean air and soil temperatures averaged for several high latitude communities in Canada. The air temperature peak and valley occur respectively a month after the summer and winter solstices. The soil temperature (one metre below ground) peaks two months after the summer solstice. It bottoms and holds steady in the second, third, and fourth months after the winter solstice. Since the snow cover insulates the ground from the warming air, and reflects much of the waxing spring sunlight back into the sky, soil temperature begins to rise only in May, three months later than the air temperature. At Resolute the soil one metre below the surface remains frozen all year (Table 6-31 in the Appendix). At high latitudes the growing season is consequently very short, usually much less than 100 days.

DEGREE-DAYS BELOW 18°CELSIUS

Every day having a mean temperature below +18°C enters the record with the difference between the mean temperature and +18°C listed as degree-days. Degree-days are tallied cumulatively for the month and for the year. Days with mean temperatures +18°C or higher do not enter the record. Degree-days offer a reliable basis for estimating heating/insulation requirements for small buildings in any location having hourly temperature records.

Figure 2-09 indicates that, strictly speaking, the heating season at high latitudes never ends. The climate rarely provides room-temperature weather.

WIND, RAIN, AND SNOW 10

TRADITION

Wind makes waves. Waves make the sea surface opaque; waves swallow kayaks. Freezing spray glazes the fishing line and the rocks along the shore. Freezing rain glazes the tundra, sometimes putting winter forage beyond the reach of muskoxen and caribou. Wind shifts the pack ice, opening impassable leads, closing them as suddenly to make impassable ridges with sails 10 metres high and keels 40 metres deep. Wind makes foam and pushes it to the lee shore of summer lakes. Wind makes the sky now mackerel, now streaked and mean, now clear. Wind makes *sastrugi*, small snowdrifts like fish scales, packed hard, eroded, and repacked in ordered chaos underfoot, always aligned upwind-downwind. Overland travellers use the lie of the drifts, day or night, to determine their direction of travel. Wind distributes and packs the stuff used to make snow houses and overnight shelters.

Wind steals heat and moisture, dispersing it aloft. Wind bounces rain off sodden tents. Wind drives snow thin as dust into hidden crannies. Wind drops mosquitoes and lifts falcons; it bends sedge stem, poppy, willow, and eyelash. Wind brings scent to hunter and hunted; it sings in spruce boughs, or snaps incoherently past the ear. In winter calm air always comes as a pleasant surprise.

Overland travellers make camp early before exhaustion sets in or the weather deteriorates without warning. A night shelter started too late invites disaster. Shelterless, the traveller can only sit, back to the wind like a sled dog, and stare down the lee tunnel of the storm. The blow may last for days. Survival depends on blood circulation and the combined insulating properties of the space suit and the snow drifting into the fur.

PROBLEM

In Rankin Inlet a small satellite dish points mutely at the ground instead of that radiation-rich spot low in the sky above the equator. A 40 kilometre per hour wind at -30°C still nags at the dish, as though trying to force it into a deeper state of unconsciousness.

In Pangnirtung, some shutters intended to protect window glass from high wind and flying objects are themselves falling from their mounts. Some restraining cables looped right over the tops of *matchbox houses* and anchored to the ground—an afterthought intended to prevent the buildings from being shifted off their foundations during wind storms—have themselves long since gone slack.

In Iqaluit, a windstorm blows the mechanical penthouse down from the roof of a highrise building. Several roofs under construction go in the same way, the wind uplifting the leading roof edges and peeling back both the new membrane and the insulation boards.

In Kugluktuk, wind-blown dust coats all the houses in the lee of gravel roads. Dust clogs the mosquito screens over the windows, reducing transparency. Dust billows into the rooms through open windows every time a vehicle passes.

In Taloyoak, a brisk north wind, already bitterly cold in August, sucks heated air out of an under-insulated, under-wrapped school; the structure shudders in response to gusts. Warm air leaks out through discontinuities between roof and exterior walls to be replaced by cold air leaking in through discontinuities between floor and exterior walls. The heating fuel metre, if the school had one, would be whizzing around.

In Yellowknife spring sunlight heats a winter's accumulation of powder snow blown into the exterior wall cavities of a highrise apartment building. Suddenly meltwater bursts from the cavities in a flood, down the interiors of partitions and along floor slabs. Three hours of frustrating clean-up later the unwelcome dam burst fades into memory to await next year's cycle.

In most northern Canadian communities the town planning grid adopted from other latitudes provides no relief from the wind. The rectangular layout funnels the wind down empty streets, often creating snowdrifts impassable to wheeled vehicles.

Wind persists almost as much as the force of gravity. It works on buildings most of the day, causing parts to vibrate and the whole to tremble. It presses on windward faces and sucks at leeward faces, imperceptibly bending the building out of shape. When wind strikes leading edges of roofs the resulting turbulence causes uplift forces sufficient to tug at the roofing, or to scour any snow, ponded water, or roofing gravel lying there. After a few years of battering, overhanging eaves begin to droop and chimneys begin to lean.

Inside the exterior wall the air barrier membrane at the upwind face of the building balloons inward—in the downwind direction—if not properly supported;

the air barrier membrane at the leeward face balloons outward—also in the downwind direction. All the problems associated with heat loss and uncontrolled passage of water vapour through the envelope ensue. Even window glass bows in and out at the whim of the wind, fatiguing the peripheral seals with lever action.

Wind cuts through the construction site, halving the productivity of the builders until wall sections, assembled on the floor deck, are tilted up to make a wind break. Wind-borne ice crystals and dust particles abrade exposed surfaces like a sand-blasting machine in a slow process of erosion, changing colours and textures. Driving rain subtly alters colours and textures through initial impact and then through freeze-thaw cycles and chemical action. Differential staining results and resilience to weathering diminishes as the chemical nature of building materials changes.

Besides weathering exposed surfaces, blowing snow reduces access and egress with snowdrifts, making passage inconvenient and increasing the risk of injury in case of fire. Large snowdrifts melt last in the spring, extending the mud season and causing more water to run into the foundation soil where surface drainage is poor. Snowdrifts change the nature of the building: the outside observer looks down into windows that in summer are above eye level. For the inside observer the view out becomes claustrophobic. At the windward edge of the Baker Lake townsite, snowdrifts bury matchbox houses up to the eaves; just the chimney pipes show through. The inhabitants must tunnel through the snow like *sik-siks* to get in and out. The houses at the windward edge act like a snow fence, accumulating snow that otherwise would overwhelm the downwind parts of the community. (In recent years tall snow fences erected upwind of the community have reduced the depth of snowdrifting around houses.)

Lee side snowdrifts.
Rankin Inlet—March

CONCEPTS

Airflow: Wind is air passing one way from a zone of high atmospheric pressure to a zone of lower pressure. Wind tends to flow in a regular stream, but on encountering obstacles near the earth's surface readily becomes turbulent. For any wind the volume of airflow upwind of an obstacle must be about the same as the volume downwind; since the path around an obstacle is longer than the imaginary one straight through it, the wind speed must increase temporarily to keep downwind air volume constant. Where the airflow splits around a building there is turbulence with a preponderance of high pressure zones, resulting in a push against the building. Where the airflow regroups downwind of the building there is turbulence around a zone of low pressure, resulting in a pull away from the building. Increase the number of obstacles on the ground by adding mountains and trees, and the air speed closest to the ground will decrease. Where there are no mountains and no trees—little surface friction—such as in the Canadian Prairies and the Canadian Central Arctic, average wind speed near the ground is comparatively high. For this reason surface winds have a greater impact on people and buildings beyond the tree line than within it. Some treeless valleys and fiords seem purposely shaped to accelerate winds to even higher speeds.

Heat loss: At high latitudes air temperature remains below the temperature of the human body (+35.6°C); body heat therefore flows outward into the air. The more fresh air there is next to the body the more heat is lost; the higher the wind speed the greater the amount of heat that the air can absorb from the body. Decreasing the temperature of the air and increasing its speed results in higher wind chill. Wind chill is expressed either as a cooling rate in watts per square metre or as the current air temperature decreased by a factor to result in an equivalent

Sastrugi. Devon Icecap—August

chill temperature. For instance, a cooling rate nomograph will indicate 2,000 watts per square metre being lost at an air temperature of -20°C and a wind speed of 40 kph; an equivalent wind chill temperature chart will indicate -45°C in calm air for the same conditions. Exposed human flesh has more heat to lose for a given surface, and less insulation to slow down the loss, than a building. Heated only to +22°C a building has less heat per unit area to lose at its exterior wall plus some insulation to slow down the heat loss rate. Human flesh—mostly liquid—cannot tolerate freezing. The outer components of a building envelope, already in the solid state (with the exception of concealed water, water vapour, and air), can withstand a certain amount of freezing and thawing. Wind chill factors commonly published refer to exposed human flesh only, not to buildings.

Rainfall intensity: Calling the vast continental expanses at high latitudes a "polar desert" because its annual precipitation does not exceed 250 millimetres—and much of that snow—is one thing. It is another thing altogether to design buildings at these latitudes as if rain were not a problem. Rain is a problem. If the mean rainfall for Yellowknife in July is 34 mm but one fine July day a local storm drops 30 mm of rain on the downtown in the space of half an hour (while the rain gauge at the airport six kilometres away registers zero rainfall), the effects of rain cannot be ignored. Wall and roof design must consider rainfall intensity, more than total annual precipitation, for rainwater to be drained away safely. The weight of water ponded on a roof stresses the building's structure. An estimate of the amount of rain to fall in 15 minutes, which will be exceeded only once in ten years, is listed for key communities in the National Building Code of Canada.

Sublimation: Ice and snow may change to water vapour without passing through the liquid phase. This direct phase change—*sublimation*—accounts for the replacement of snow flakes by ice crystals and air spaces at the bottom of the snow cover. In winter the wind picks up vapour from the snow-covered ground, increasing the relative humidity of the air and the potential for precipitation.

Snow cover: Since snow cover contains large amounts of still air and still air is a good natural insulator, snow cover insulates the ground, sea ice, and even roofs against the cold. Snow cover prevents the outside winter air from deep-freezing the earth surface or the sea water below the ice. With outside air at -40°C the ground surface temperature may be only -15°C and the sea water -2°C. Increasing the thickness of the snow cover raises the temperature of the substrate. For generations people inhabiting leaky buildings have packed snow around the foundations both to raise the temperature of the crawl space and to reduce the rate of air infiltration through the floor.

Snow and wind: Air moving as slowly as 10 kph can lift the lightest snow and ice particles at the top of the snow cover and blow them downwind. As long as an upwind supply of snow and a 10 kph or faster wind co-exist there will be a play of snow deposition and erosion downwind. Open tundra upwind provides an inexhaustible supply of snow; for various reasons boreal forest, open water, and wind-swept mountains do not. Although individual crystals may last only a kilometre or two in the airstream before disintegrating, fresh particles leapfrog the old to keep the supply steady. Obstacles in the airflow cause local turbulence. Wherever there is turbulence pockets of air move at speeds lower than that of the main airstream. Whenever air speed drops below the minimum, snow particles drop out of the airstream to form snowdrifts.

MEAN WIND SPEED

Flat, treeless terrain has higher wind speeds annually than typical points in the mountainous regions of western North America. The high mean wind speeds of Cambridge Bay, Baker Lake, and Resolute—22 kph—are comparable to those in Regina, Winnipeg, and Sudbury (Table 9-31 in the Appendix). At high latitudes mean wind speed hardly varies all year. Glaciated highlands at high latitudes produce chinook winds that, joined with funnelling topography, make the freakish windstorms that shake some buildings from their foundations. Several communities in northern Canada have recorded gust speeds greater than 150 kph.

Wind speed by itself only tells the building designer what impact forces the building envelope must resist. The prevailing wind direction, winter and summer, must be known in order to ascertain which sides of the building will be most prone to heat loss and solar heat gain, to wetting, and to snow-drifting; and to estimate wind behaviour in the spaces between the building and its neighbours. Wind direction data are available for many high latitude locations (Table 10-31 in the Appendix). Read with mean snowfall data (Table 13-31 in the Appendix), wind direction data can indicate the direction most likely to spawn snowstorms accompanied by snowdrifting.

A wind rose that combines wind direction frequency and speed for the locality is essential for the orientation of buildings and the placement of windows and entries. A wind rose such as the one shown in Figure 1-10 can be compiled from the statistical data published by the Atmospheric Environment Service of Canada. Since weather stations are often located at airports, wind direction data may not hold

Transfigured sky over Truelove Lowland. Devon Island—July

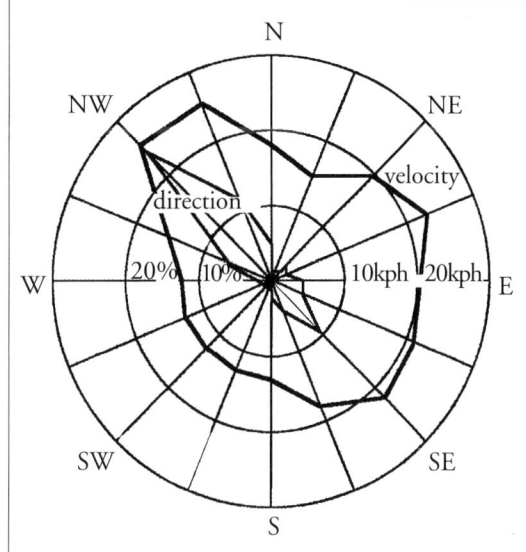

Figure 1-10 Wind direction frequency (%) and mean velocity (kph) at Iqaluit for the months of September to May.

true at the townsites. Local topographic effects and certainly the shape of the village plan influence air flow. In the absence of published data for a given location information must be sought from people living in the village. Careful interpretation of snow-drifting patterns within the community may give a reliable fix on the wind directions prevailing in winter.

WIND CHILL

The colder the air next to skin the higher the rate of heat loss from the human body. In calm air with the body at rest the thin layer of cold air stagnating against the skin will warm up slightly. But if the surrounding air moves constantly, new cold air replaces the warmed layer and the heat loss rate spirals upward. Table 10-31 in the Appendix indicates that typical wind chill rates in Northern Canada are almost half again as high as those at midlatitudes.

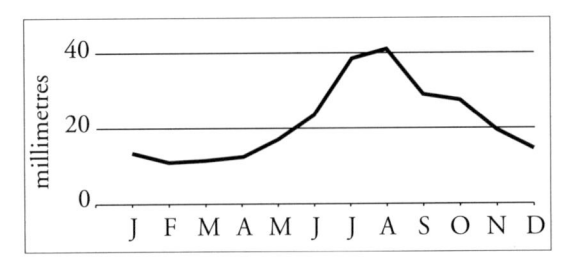

Figure 2-10 Total precipitation (in millimetres), averaged for eight northern communities in Canada.

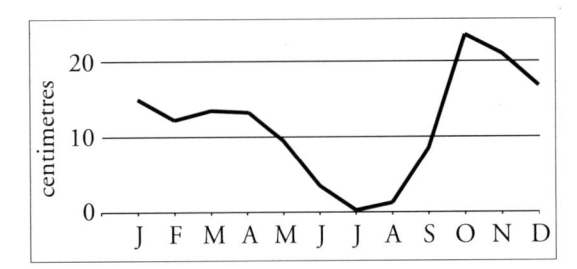

Figure 3-10 Mean snowfall (in centimetres), averaged for eight northern communities in Canada.

Clothing design must compensate appropriately, since exposed flesh freezes readily at chill rates above 1,600 watts per square metre. This makes careless exposure at Resolute, for instance, a high risk from October to April—seven months of the year.

TOTAL PRECIPITATION
Except in east and west coast areas influenced by local maritime climates, little rain and snow falls in continental areas at high latitudes. Annual totals less than 250 mm are common. Polar air, being comparatively very cold, has little vapour-carrying capacity in absolute terms. Not much vapour means not much rain or snow. Figure 2-10 indicates that precipitation peaks in summer after snow melt and ice break-up. Rain, usually drizzle, may last for hours and sometimes days at a time. High wind speed, local turbulence, and drizzle combine as driving rain. Under such conditions no vertical surface of a building is safe from frequent wetting. Driving rain will enter the building envelope where an effective air barrier system has not been provided. Thunderstorms beyond the tree line are rare. Hail is almost non-existent at high latitudes in Canada.

Rain is not pure water at any latitude; reaction with atmospheric gases and pollutants creates a dilute acidic mixture with consequences, over time, for the surface durability of exposed materials.

MEAN SNOWFALL
At high latitudes about half of annual precipitation falls as snow; at midlatitudes snowfall amounts to only a fifth of the total. This means that the high and midlatitudes of Canada receive about the same amount of snow every year, the difference being in distribution over the year—nine months with snowfall versus six (Table 13-31 in the Appendix). Figure 3-10 shows that peak snowfall at high latitudes occurs in the fall before the large expanses of open water freeze. Little snow falls when air temperatures are very low. Snow particles are usually grainy, although large hexagonal crystals are common within the tree line.

BLOWING SNOW
Distinct from snowstorms, blowing snow refers to airborne snow particles that reduce horizontal visibility at eye level to 10 kilometres or less. It is densest in the first 10 metres above the ground surface. It causes *whiteout*—loss of horizon and the disappearance from view of people and buildings. It deposits fine powdered snow in any building cavity having a "back door" for the passage of air. Blowing snow causes both snowdrifting and drift erosion; it also abrades exposed materials. Distribution of snow cover over the ground varies greatly. Depressions in the terrain, forests, and the lees of obstacles large and small fill with snow. Flat expanses and ridges exposed to the wind do not.

SNOWDRIFTING
For severe snowdrifting to occur a large upwind supply of snow, a strong prevailing wind over smooth ground surfaces, or a major snowstorm must co-exist with one or more local obstacles. Although annual snowfall at high latitudes is slight, the land's huge storage capacity guarantees supply. In between ice ages—the present era for instance—snow supply is interrupted once a year by above-freezing air temperatures, which first bind the surface snow, then melt it. Without warm summers the majority of buildings at high latitudes would have long since disappeared under a mantle of snow and ice. There are seasonal variations too—some winters go down in local history as being snowdrifting monsters. The snowdrifts of intervening years seem tame in comparison.

Even within the tree line snowdrifting affects areas exposed to the airstream for any reason. A typical school, located in a cleared area surrounded by trees, has a gymnasium whose mass stands higher than the classrooms. Most winters snow drifts heavily onto the low roof at the leeward side of the gymnasium.

11 HUMIDITY

TRADITION

In winter the overland traveller fears wet clothing more than game hovering beyond weapons range or chance encounters with bears. Staying dry is essential to staying warm. Clothing soaked by a fall through thin ice freezes instantly at low air temperatures, putting recovery beyond reach. Even the routine build-up of perspiration inside a space suit made of animal fur can lead to hypothermia if left unchecked. Animal bodies at work generate waste heat. Human bodies perspire to lose some of the heat by evaporation. The resulting water vapour migrates through the fur of the space suit toward the zone of lower *moisture content* outside. On contact with cold surfaces in the outer layers of the space suit the vapour condenses, wetting the material. It then freezes as soon as the waste heat dissipates. When water, a poor thermal insulator, invades the spaces in the fur that contain still air, a good thermal insulator, the space suit loses much of its insulating value and the body inside starts to chill.

Inside snow houses relative humidity remains high, the air temperature hovers around freezing, and lung ailments head the list of common illnesses. This is not surprising given that snow houses are made of water in the solid phase. However, as in any other shelter, the warmest air inside rises to the ceiling. Damp clothing hung near the ceiling will eventually dry out. Water vapour that condenses as frost on the exposed hairs of fur clothing can be knocked off with a beater designed for the purpose, just as a sled dog shakes its body to remove frozen breath from its fur.

PROBLEM

Designers of high latitude buildings must control all phases of water, whether ice (frost), water (leaks), or water vapour (humidity), in order to avoid ice jams in doorways, inside spaces soaked by melted frost leaked from attics, or human skin as brittle as parchment. Inside the structure water will corrode metal and rot wood. It will soak building insulation, making it more useless than wet clothing, since buildings cannot be hung out to dry. Nor can frost be beaten out of building insulation.

Consider a house with a cathedral ceiling (no attic space), untreated wood for roof joists and roof deck, asphalt shingles for roofing, and a polyethylene vapour barrier on the inside covered by plasterboard. It has a fluorescent light fixture box recessed into its roof-ceiling assembly, causing a break in the vapour barrier. (The construction described here contravenes the National Building Code of Canada in significant ways, but codes alone do not prevent errors.) The warmest air in the house rises to the ceiling and some escapes into the roof-ceiling assembly through the joints around the fixture. The water vapour contained in the escaping air condenses on contact with the colder air and surfaces in the joist space just below the roof deck. Since the roof deck and the roofing itself resist vapour migration, the water and water vapour that accumulate here cannot escape to the outside. So in the joist space around the fixture the essential conditions for wood rot co-exist—wood with little resistance to bacterial growth, trapped water, oxygen, and temperatures warmer than +5°C provided by the lamp fixture. This section of roof can be expected to collapse long before the rest of the building.

It is just as much a problem to have too little water vapour inside a building. Structural wood that has high moisture content when installed will shrink severely in a dry environment, opening joints and starting cracks in both the building envelope and the finishes. Very dry air irritates the eyes, inflames the nose, chaps lips, parches the throat, cracks skin, and splits fingernails. The stress on lung surfaces can induce respiratory illness. In northern Canada artificial humidification is applied inconsistently. Many buildings have none. A person's home may be humidified to 60% relative humidity (rh) in winter (making the windows opaque with condensation), while the humidity at his or her workplace never exceeds 15% rh. Moving repeatedly from one environment to the other stresses body tissues.

The problem of humidity in buildings is not just technical. A wide gap between the physical reality and most people's understanding of it persists, partly because the physics of moist air involves many variables. (Air temperature has a single variable; wind speed has two, distance and time.) An increase in the number of variables leads to confusion. Intuitively humans realize how humidity affects their lives—some talk of preferring "dry-cold" winter climates (many sunny days) to "wet-cold" winter climates (many cloudy days). Others note the extreme dryness of some buildings in winter as wood furniture develops cracks, static electrical sparks fly up from the carpet, and skin and hair become brittle. In summer

CLIMATE

*100 percent relative humidity.
Devon Icecap—June*

some humans react uncomfortably to "sticky heat" and declare a preference for "dry heat." Many people view humidity with misgiving as a result, conceding that its control remains as out of reach as the weather itself. Add to this a long list of technical terms including synonyms and near synonyms (psychrometry, moisture content, *vapour pressure*, relative humidity, dew point, saturation, wet bulb temperature, enthalpy, psychrometric chart, air volume, barometric pressure, permeance, sorption isotherm, to name some), and the layman's wariness regarding the study of humidity seems almost healthy. Still, a safe, comfortable building cannot be designed for high latitudes without understanding the basics of the air-vapour mixture.

CONCEPTS

Water vapour is an invisible, tasteless, odourless gas. It is not steam, fog, mist, or cloud—these are visible concentrations of water droplets that have condensed from water vapour. It is a small component (typically much less than 2 percent by mass at high latitudes) in the mixture of gases making up the air of the lower atmosphere.

The ratio of the mass of water vapour present in an air sample to the unit mass of air with no vapour present is called the moisture content. It is expressed as kilograms of moisture per kilogram of dry air (kg/kg Dry Air). Moisture content must be considered when mixing cold outside air with the air of heated interiors.

The term relative humidity (rh) refers to the ratio of the actual moisture content of a given parcel of air to the moisture content of the same parcel if it were saturated at the same temperature. It is usually expressed as a percentage. After air temperature relative humidity (not moisture content) is the measure most relevant to human comfort in a heated room, and to the "comfort" of building materials (wood, paper, concrete, steel) that react dimensionally or superficially to changes in the interior air's relative humidity. At a constant temperature relative humidity varies when moisture is added or subtracted from the air, as might be expected. But at constant moisture content it also varies downward when the air temperature rises, and upward when the air temperature drops.

In a given parcel of air the relative humidity can vary from zero up to a natural limit (saturation, or 100% rh) imposed by the temperature and barometric pressure of the sample. (The influence of barometric pressure on moisture content is slight for most building purposes and will be ignored here.) Warm air can contain much more water vapour at saturation than cold air. For instance, the moisture content of -30°C air at saturation is 0.00025 kg/kg Dry Air—next to nothing. The moisture content of +23°C air at saturation is 0.018 kg/kg Dry Air—about 70 times greater.

The low moisture content of very cold air at saturation compared to warm air is crucial to high latitude building design, since some cold air must be brought in to ventilate heated buildings in winter. But to be useful the outside air must be heated to room temperature. Heating the air does not raise its moisture content, just its potential for holding water vapour. Table 1-11 shows what happens to relative humidity when an outside air sample at -25°C and 80% rh is heated to +23°C (room temperature): the moisture content remains constant but the relative humidity drops to 3%. The result is very dry air, air with low relative humidity, air that has plenty of spare capacity to absorb moisture from surrounding materials. This is akin to replacing a small kitchen sponge that is nearly saturated with a much larger one containing an identical amount of water. The water is lost in the larger sponge, leaving it relatively dry and capable of absorbing much more water.

Lowering the temperature of the air to its dew point or adding moisture to saturated air will cause vapour to condense as water droplets (dew, condensation, mist, fog, cloud, steam), or as ice crystals (hail, snow, ice fog, frost) if the temperature of the sample dips below freezing point. A kitchen sponge filled with water to saturation will expel any additional water poured onto it. Exhaling deeply into a freezer produces an ice fog: the moisture content of the breath greatly exceeds the moisture content of the below-freezing air at saturation. Table 2-11 shows what happens to room air at +23°C and 20% rh when it leaks out of a building in winter. The air cools rapidly and its relative humidity soars to saturation long before its temperature drops to the level of the ambient air. Condensation occurs within the thickness of the wall, but not on inboard wall and window surfaces whose temperature remains higher than the dew point of the room air. Table 3-11 illustrates the case of room air with twice the relative humidity. Saturation is reached at a temperature higher than window temperature—water vapour condenses on the glass, the coldest surface in the room. The temperature at saturation is termed the dew point temperature for the air-vapour sample.

The force of gravity makes water "find its own level." Because water vapour is an independent gas in air, it diffuses through the air molecule by molecule to establish a moisture equilibrium. Given two adjacent zones with a difference in moisture content water vapour moves from the zone of high moisture content to the zone of low moisture content. In a heated, inhabited building at high latitudes the moisture content of the air inside usually exceeds the moisture content of the air outside in winter, so the water vapour inside tends to diffuse outward through the building envelope. The steeper the difference between inside and outside moisture contents, the quicker the diffusion of water vapour between the two zones.

Air currents also move water vapour from one place to another. Wind pressure, *stack effect*, and mechanical ventilation systems propel air through building elements toward zones of lower air pressure.

Table 1-11 The steep drop in relative humidity when outside air at -25°C and 80% rh is heated to room temperature without changing moisture content.

Zone or transition	Temperature °C	Relative humidity %	Moisture content kg/kg DA	State
Body	+36	1	0.0003	extremely dry
Room	+23	3	0.0003	very dry
Wall (RSI 5)	+22	3	0.0003	very dry
Window, double	+5	6	0.0003	dry
Ice melts	0	8	0.0003	dry
Outside	-25	80	0.0003	very humid
Dew point	-28	100	0.0003	saturation

Table 2-11 The steep rise in relative humidity when room air at +23°C and 20% rh is cooled to outside air temperature without changing moisture content.

Zone or transition	Temperature °C	Relative humidity %	Moisture content kg/kg DA	State
Body	+36	9	0.0035	dry
Room	+23	20	0.0035	dry
Wall (RSI 5)	+22	21	0.0035	dry
Window, double	+5	65	0.0035	humid
Water freezes	0	90	0.0035	very humid
Dew point	-1	100	0.0035	saturation
Outside	-25	100	0.0035	frost, ice fog

CLIMATE

Table 3-11 The steep rise in relative humidity when room air at +23°C and 40% rh is cooled to outside air temperature without changing moisture content. Water vapour condenses on window surfaces.

Zone or transition	Temperature °C	Relative humidity %	Moisture content kg/kg DA	State
Body	+36	19	0.007	dry
Room	+23	40	0.007	moderately humid
Wall (RSI 5)	+22	43	0.007	moderately humid
Dew point	+9	100	0.007	saturation
Window, double	+5	100	0.007	water droplets
Water freezes	0	100	0.007	frost
Outside	-25	100	0.007	frost, ice fog

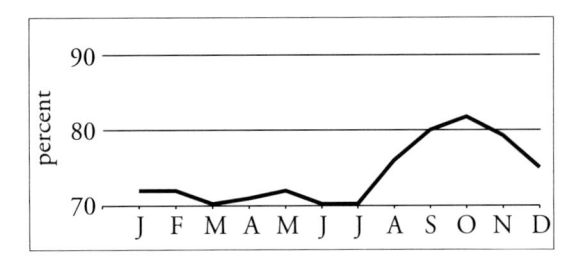

Figure 1-11 Relative humidity, averaged for eight northern communities in Canada.

Since air currents have the potential to replace the air contained in a building up to several times an hour, the potential for moving the water vapour contained in the air is great. Over a period of time much more water vapour will travel out of a typical building in air currents than could be displaced by molecular diffusion. Given that the moisture content of the air inside is greater than that outside, more water vapour leaves the building with air currents than enters by air infiltration.

Compared to dry air and to many building materials, water vapour, like water, has high heat-storage capacity. It takes about twice as much energy to heat water vapour as dry air. Therefore, any water vapour lost to the outside through molecular diffusion or air leaks represents a loss of heat from the building as well.

SOURCES

At high latitudes an abundance of surface water, ground water, snow, and ice enters the lower atmosphere as water vapour through evaporation, transpiration, and sublimation. Snow and ice sublimate throughout the winter, resulting in average relative humidities around 75% rh, as though contradicting the geographic designation "polar desert." (But given the low vapour-carrying capacity of cold air the moisture content of outside air at high latitudes does remain low.) Inside buildings, human lungs, perspiration, laundry, cooking, and bathing contribute water vapour. Freshly installed wet materials such as

85% relative humidity. Broughton Island—May

plaster, mortar, concrete, and latex paint contribute water vapour while they set and dry out. Portable heaters (salamanders) used in winter construction contribute water vapour as a by-product of combustion.

Figure 1-11 shows mean relative humidity month by month, averaged for eight northern communities in northern Canada (see also Table 15-31 in the Appendix). In northern Canada it peaks in the late summer and early fall, when the lake and sea ice have broken up, exposing vast expanses of water to evaporation in the cooling atmosphere. This means that more days with fog occur in the fall at high latitudes—a nuisance to sealift operations and air travel—than at any other time of year (Table 17-31 in the Appendix).

EFFECTS

How does humidity affect us? Why do we experience "dry cold" at high latitudes when it is so relatively humid there? What counts for human comfort is not the relative humidity in the atmosphere around us so much as the relative humidity of that thin layer of air closest to our bodies. In winter this thin layer is much warmer than the surrounding air for obvious reasons, so its capacity for carrying water vapour is greater. Table 1-11 indicates that body heat converts the cold, relatively humid air around us into warm relatively dry air. The moisture content remains unchanged. This explains why humans and materials containing moisture dry out so severely inside heated buildings at high latitudes. Of course, humans can compensate by drinking more fluids, but some compensate less well than others.

Many materials cannot compensate for dryness at all. Wood contains moisture. It shrinks when its internal moisture content drops below about 30 percent by mass. Below 20 percent it shrinks much more, and splits and splinters. Static electricity, generated by the friction between shuffling objects and ordinary carpets, is normally grounded by water vapour in relatively humid air. Water is a good conductor of electricity. Dry air is a good insulator of electricity. The static electricity accumulating in a shuffling object surrounded by relatively dry air has nowhere to go until the object makes contact with another conductor, such as a door handle made of metal or the metallic parts of an electric light switch.

Buildings in winter dry out as a whole when outside air, cold and relatively humid, is heated to room temperature. The air's moisture content remains unchanged but the relative humidity drops precipitously.

Damage results from the condensation of water vapour on surfaces whose temperatures fall below the dew point temperature of the inside air. The cycle of wetting and drying causes staining and mould growth followed by disintegration.

IV
PROGRAM

12 SETTLEMENT
CLOSE AND FAR *61*
TRADITION *61*
TRADING POSTS *62*
MISSIONS AND POLICE *63*
SCHOOLS AND NURSING STATIONS *63*
PUBLIC HOUSING *64*
HERE AND NOW *64*
FUTURE CONSIDERATIONS *65*

13 SHELTER
TRADITION *66*
SPACE SUIT PRINCIPLES *66*
WINTER SPACE SUIT CONSTRUCTION *66*
SYMBOLIC CONTENT *67*
SELECTION OF THE FITTEST *67*
ATTRIBUTES OF THE SNOW HOUSE *68*

14 OWNER
BAD BUILDINGS *71*
INSTRUCTING THE ARCHITECT *71*
CULTURE GAP *72*
CULTURE BRIDGE *72*
PROJECT IMPACT *73*
JURISDICTIONS *74*
PROJECT COSTS *74*

15 LOGISTICS
TRADITION *76*
PROBLEM *76*
PROCUREMENT *76*
PACKAGING *76*
LAND TRANSPORT *77*
WATER TRANSPORT *77*
AIR TRANSPORT *78*
STORAGE AND ACCOMMODATION *78*
EQUIPMENT *78*
COMMUNICATION *79*

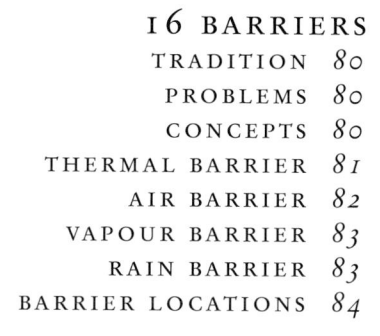

16 BARRIERS
TRADITION *80*
PROBLEMS *80*
CONCEPTS *80*
THERMAL BARRIER *81*
AIR BARRIER *82*
VAPOUR BARRIER *83*
RAIN BARRIER *83*
BARRIER LOCATIONS *84*

Kimmirut—April

60 PROGRAM

Nuuk—August

Zoning plan—Igloolik

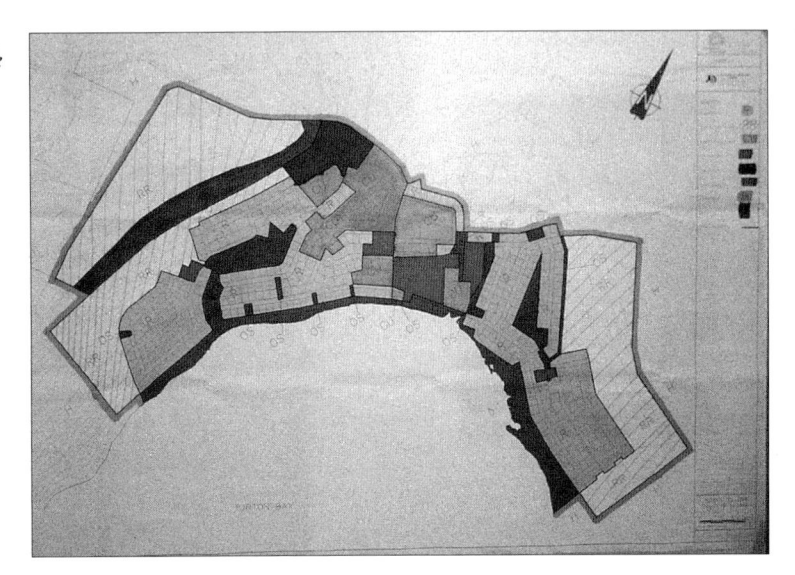

SETTLEMENT

CLOSE AND FAR

Manhattan on a clear night, seen from twelve thousand metres up, is breathtaking. Skyscrapers full of parcelled light grow skywards from the darkness as though etched on a silicon memory chip. Manhattan seems to be pure intelligence, a presence greater than the sum of its parts, proof that to humankind belongs the same majesty, the same complexity, the same momentum that humans attribute to other processes in nature. Seen from a grassy rise higher than the cottonwoods lining the river bed, the encampment of a hundred conical lodges on the opposite bank of the Little Big Horn River conveys the same message of collective, purposeful intelligence and endurance. A reddish glow seeps through the skin of each tipi, wavering like starlight, and a hundred smokes rise in the still air. Seen from the fiord, the houses of west Greenland villages perch on outcrops like seabirds at rest, confident and vigilant. Each house is rooted in the rock, its roof peaks like the rock, and each has a view of the sea and exposure to the sun.

These places radiate the self-assurance of virgin forests, watering elephants, and lava volcanoes. Clearly such places, reconnoitred on foot, reveal a web of detail as inscrutable as the microscopic dots that make up a magazine photo. Ambition, clairvoyance, productivity, social organization, and co-operative behaviour vie for ascendancy with skewed hopes, destitution, despondency, anti-social behaviour, self-righteousness, conspicuous consumption, and suicide. While such turmoil runs through all societies, some, unaccountably, develop shelters and urban structures that belie the chaotic energy found just below the hard surfaces. A self-assured whole develops, greater than the sum of its parts.

Most forty- and fifty-year-old communities in northern Canada do not radiate such assurance.

TRADITION

In the centuries preceding permanent settlement in northern Canada seasonal and semi-permanent camps did radiate self-assurance. The people living in them chose their building sites intuitively, making the best of formations and materials offered by familiar landscapes. Impermanence made site improvements possible the following season; in the off season, wind, snow, and rain freshened the site, returning it to inventory.

The first criterion for selecting a seasonal campsite was being in the right place at the right time to harvest fish and game in abundance—a fish run, a caribou crossing, sea ice peppered with breathing holes—and being in tune with the spirits. The knowledge of such places, times, and spirits was kept in the heads of elders. The occasional failure of knowledge led quickly to deprivation.

Second, waterfront access to the means of travel—river, lake, seashore, or sea ice—was essential. Waterways led directly to the hunting ground and to the next campsite. The foreshore had to be high enough above tide or flood level for secure storage of canoes or kayaks, but sloped and low enough to launch boats and haul out carcasses. Places with congested ice conditions were avoided.

Third, the campsite required a view of the most likely land and water approaches so that camp sitters could keep an eye out for returning hunters, sign of game, abrupt changes in weather, and the arrival of enemy raiders.

Fourth, the site had to be well drained, with enough space for four or five families in separate tents or houses, plus drying racks, smoke houses, storage platforms, or meat caches. It had to offer some building materials—trees for making poles, or whale ribs for roof structures, or flagstones for floors and platforms.

Fifth, the site had to be suitable for orienting shelters with their backs to the wind and their fronts to the sun at the equator. Predators investigating camp scents from downwind came into view before they could do damage. Broken ground (gullies, cutbanks, pingoes), favoured by bear and wolf for denning, was avoided.

Sixth, a convenient supply of drinking water was required. Where fire was practicable, the area surrounding the site had to offer firewood in the form of deadfall, driftwood, or dwarf willow.

The hunting community was small, consisting of a few extended families, the maximum size that could be sustained by a hunting ground only two or three day's walk in breadth. Wherever game was especially plentiful the assembly of shelters grew larger. Tents or houses were placed three to ten metres apart in a cluster, so that members of extended families remained in close proximity. Sometimes tunnels between adjacent shelters were built to make visiting

Repulse Bay—March

easy. At large gatherings the shelters were strung out along the beach or river bank so that all had equal access to the waterfront and relief from the neighbour's campfire smoke.

TRADING POSTS

The steady evolution of hunting/gathering cultures in northern Canada hit a snag in the eighteenth century when immigrants from the Old World, keen to extract furs for the clothing market in Europe, established trading posts on major waterways. To the surprise of the indigenous peoples, the winter fur of small animals previously considered tedious to catch and poor for eating could be exchanged for exotic goods such as glass beads, metal, tea, sugar, flour, tobacco, guns, and ammunition. The difficulty remained that trapping was a life's work in itself. Garnering enough white fox pelts to sell to the trader left little time to hunt game for food. Trapping grounds did not conveniently overlap hunting grounds, especially as areas quickly became trapped out, forcing moves to ever more distant locations.

The choices—living off the land and trusting to fate or living off the fur-bearers and trusting to trader logic—probably never seemed mutually exclusive. Why not have both? A hybrid form of settlement resulted. People hunted part-time, trapped part-time, and camped around the trading post part-time. The post, having stood alone in the landscape, became the core of a seasonal shanty town.

To be successful the new post had to be located in the midst of hunting and trapping grounds; animals, and people to capture them, had to abound. It made no sense to locate where there was one without the other. With sketchy maps in hand the trader had to reconnoitre coastlines by boat during the brief shipping season. Mistakes were common. Some posts lasted as little as a year, having been located in the wrong place; they were dismantled and rebuilt in another location suggested by better information.

Second, the site had to be accessible by sea or river to the annual resupply ship. The trader had to judge that the ice conditions of late summer encountered in the first year would be typical for subsequent years. Without resupply the trading post had no basis for being. The site had to offer a sheltered anchorage for the resupply ship and a gentle foreshore suitable for landing its barge, or a channel out of the wind and deep enough at low tide to allow shoreside mooring.

Third, the building site had to be in sight of both anchorage and landing, flat or gently sloped, and well-drained to support above-ground buildings made of wood for many years. The space around the post had to be suitable for receiving temporary camps of trappers and their families gathered to trade furs for food and ammunition.

Fourth, a nearby stream or lake to supply water or ice blocks for drinking, cooking, and washing was essential.

The trading post usually consisted of the post itself, about ten by twelve metres in plan, a house with several bedrooms, a warehouse, a coal shed, and a powder shack, all clapboard on wood frame. The Hudson Bay Company's buildings were painted white with green trim and had red gable roofs. Only the house was heated. Trading in winter was done under a cloud of condensed breath in the light of oil lanterns; hands remained flexible for half an hour at a time. Hot tea served as antifreeze.

At a successful trading post traders and trappers recognized the dependence of one on the other. If the hunting or trapping in the vicinity failed, the trader would assist the trappers with food and the means to relocate in a better area. If either was stricken by disaster the other would provide assistance. The trader, equipped with a book of diagnoses and remedies, was the closest thing to a medical doctor available. He usually hired a local assistant or two to help with the business of storing, shipping, and trading. Despite the trader's lack of interest in founding a village around his post—he wanted people out on the land harvesting fur—and despite differences in culture and lifestyle, this intense collaboration was the seed for permanent settlement.

MISSIONS AND POLICE

The evangelists and the police preferred to see permanent settlements established before hazarding missions and stations in the remote regions of northern Canada. Anglican and Roman Catholic missions needed a convertible congregation within sight and sound of church bells; the police needed a compact population to serve and protect without incurring exorbitant travel costs. The traders were the advance party: if a community could coalesce around a trading post and survive for two or three years at least, the bishop and the commander sent in their own advance parties.

Each mission or station built a cluster of small buildings: institution, housing, and storage. These, added to the buildings established by the trading company, gave at least the appearance of a permanent settlement. The physical planning was largely intuitive. The buildings were sited in accordance with the mission leader's own criteria, independent of the others present and of any notion of planned development. They did share a need for a convenient water supply and a downstream location for sewage disposal.

Co-operation between the first Euro-Canadian inhabitants was pragmatic rather then formal: each group's leadership and code of behaviour was an extension of standing orders long established in a far-off society. These outposts worked as branch offices. So the clusters tended to stand off from each other. In the spaces left between grew the native's campsites, often strung out parallel to the water so that each extended family could remain together while having similar access to the shore, sunlight, and views. The settlement had hard parts for housing Euro-Canadians and their institutions and soft parts—tents and shanties built from packing cases—for the natives. The latter still spent most of the year on the land hunting and trapping, fuelling the settlement's economy.

SCHOOLS AND NURSING STATIONS

Released from a long European war and confronted with reports of disease and destitution among northern native populations, the Canadian government mobilized itself in the 1950s to bring support services to remote settlements. People with infectious diseases were flown to urban centres in southern Canada for treatment. Nursing stations were built to monitor health and dispense disease prevention technique. Family allowance and welfare cheques were issued to supplement income. Children were sent out to northern towns for schooling until a local elementary school could be built and staffed. Acceptance of these services meant acceptance of other major changes—for instance, the threat of stopping family allowance cheques convinced many families to leave their life on the land so that children could attend school during the winter months.

Rankin Inlet—April

This deep incursion in the accepted way of living on the land had predictable effects. Children sent to outside schools came back with intimate knowledge of how to survive in a hostile, urban environment. They were forced to learn the basics of a foreign way of life in a foreign language of instruction. The languages of their ancestors, deprived of new generations of speakers, began to die. They also learned that parents and elders in urban societies do not enjoy the universal respect essential to survival on the land.

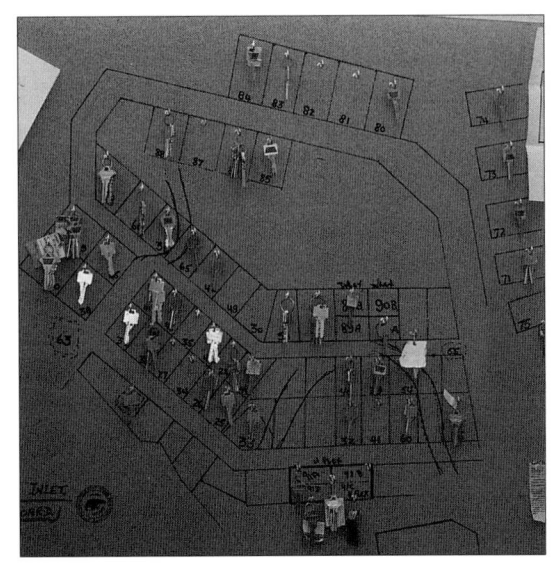

Key board at housing office—Chesterfield Inlet

With the school and the nursing station came diesel power generation—an uninterrupted noise and smell that underscored the separation of village life from life on the land. New conveniences and new rules turned the camp around the trading post into a permanent settlement. Buildings so large had no precedent: people living next to them in tents were dwarfed. Big buildings changed the landscape, blocking the view, throwing long shadows, funnelling the wind, and creating huge snowdrifts, all the while making enormous energy and service demands until then unimagined.

PUBLIC HOUSING

If people were obliged to spend the winter in the settlement so that their children could be schooled and inoculated, how were they to be housed? Drafty tents and damp snow houses occupied for longer than normal periods would make a mockery of public health ambitions. The government erected houses, modest in quantity, size, and utility. To this day supply has not caught up with demand. Many people have spent a fifth of their lives wait-listed for housing. When a house became free it was allocated to those most in need and longest on the list. The old pattern of proximate co-existence among members of the extended family dissolved. Grandparents might end up living in a house at the opposite end of the settlement from their grandchildren, nieces, and nephews. It became increasingly difficult in practice to share labour, food, and good times. The self-conscious nuclear family began to supplant the collective consciousness of the extended family.

New houses in turn demanded delivery of services—electricity for light and cooking, fuel oil for heat, tanks and piping for water and sewage—in environmental conditions of a severity never studied in the technical schools of southern Canada. For the graduates of such schools and their government employer the task of implementing housing and services in remote northern settlements became a technical challenge having little reference to traditional patterns of settlement. The technicians filled the apparent vacuum of local determinants for planning permanent settlements with the determinant most familiar to themselves, the efficient distribution of utilities. The planning of utility networks dictated the planning of the community. Compact utility networks were cheaper to build and operate than extended ones. Only the fire control requirement for minimum separation between buildings (12 metres) prevented even greater density.

Governments imposed social structures and physical developments in spite of, rather than in harmony with local conditions. In extreme cases they declared a few existing communities obsolete. Fort Rae was too low and swampy, so Edzo was built ten kilometres away on higher ground. Aklavik was too prone to flooding and erosion, and was too isolated, so Inuvik was built. But the new settlements, planned on the basis of imported ideology, failed to attract all the inhabitants of the old. So now, four separate communities exist where there were two.

For a few families in each settlement the incursion of outside ways and imperatives was too much to tolerate. They left the settlements to spend most of the year in "outpost" camps, tiny shanty towns located at traditional hunting and fishing spots, to live out their lives to the rhythm of the seasons.

HERE AND NOW

For the first time in their long history native communities in northern Canada and Greenland were living beyond their means—and that at the behest of well-intentioned outsiders. People once fully employed on the land became chronically unemployed at the stroke of a pen, and were then tarred with the stigma reserved by outsiders for people on welfare. In the last quarter of the twentieth century the European demand for fur declined abruptly. Poverty in the settlements risked being permanent rather than temporary. The social consequences of such radical change will be felt well into the twenty-first century and are too numerous to list here. Instead, consider some of the physical consequences of present community planning.

People accustomed to seeing and hearing the land that supported them are shut into village plans reminiscent of the bedroom communities proliferating around southern Canadian cities. Planners locate houses in the centres of the smallest possible lots consistent with minimum fire separation, so that each house blocks the view of its immediate neighbours. Window orientation in houses ignores views and preferred exposures to sunlight.

Whereas beach ridge or riverfront served as community highroad for pedestrians and dog teams, the switch to heavy wheeled vehicles for water, sewage, and fire service entailed construction of roads. Poorly built with scarce material, equipment, and capital, the roads interrupt surface drainage in the spring, and in summer provide a steady supply of mud to be tracked into buildings. On dry summer days dust blows into the living spaces. In winter, money must be spent to keep roads clear of snowdrifts so that the wheeled service vehicles can reach every building.

Beyond the tree line, a string of houses laid out along a beach ridge lying perpendicular to the prevailing wind throws a shallow snowdrift harmlessly onto the foreshore. The addition of more houses upwind of the string makes the village act like a snow fence designed to precipitate snowdrifts. The houses on the upwind edge of the village disappear in snowdrift up to the eaves.

FUTURE CONSIDERATIONS

The primary lesson that indigenous peoples want outsiders to learn is that decisions made for natives by outsiders acting alone are useless and unacceptable. Such decisions, by definition uninformed, can only do harm. Present-day governments sometimes insist that planners in northern Canada involve inhabitants in the physical planning of existing communities. Although remote communities increasingly demand and welcome such co-operation, the result remains that the fundamental choices have already been made by outsiders and that the comparatively minor choices remaining come from the same textbook.

A community plan must contain a vision of the future conceived by its inhabitants and a flexible, responsive process for making large and small intermediate decisions. At present, existing plans were not conceived by the people most affected, nor are they flexible or responsive.

A credible process for planning the community is an evolution of self-image. It does not move or remove existing structures overnight, but makes logical decisions over a period of one or two generations that cumulatively amount to a total revision of the past. It depends on the maturity of all participants to see the burning issues of the day in the broader context of their children's future. More importantly, it depends on local society knowing itself, recognizing and understanding both its current position and its preferred way of living. No society out of touch with itself can create viable human settlement.

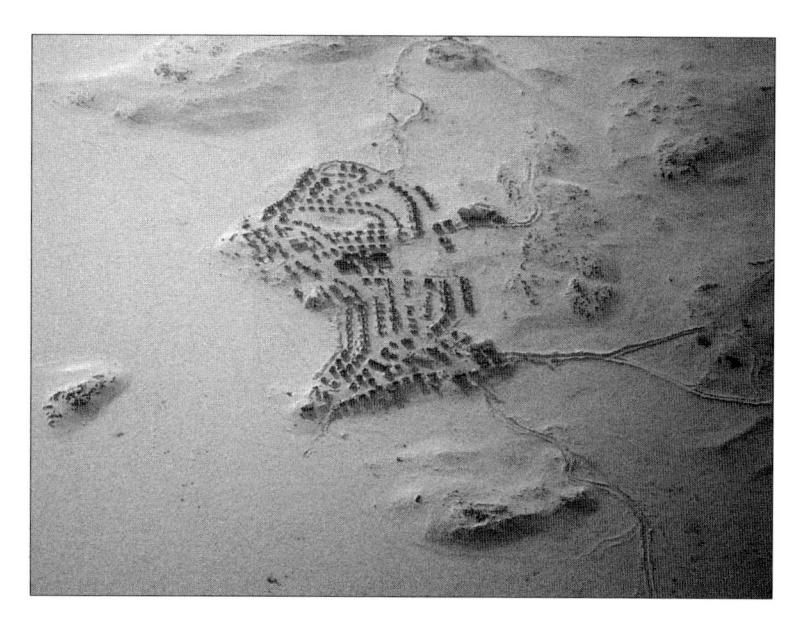

Village plan seen from the air. Povungnituk—April

SHELTER

TRADITION

Little used to the land, midlatitude city dwellers wear windbreakers and live in climate-controlled space modules. Inuit used to living continuously on the land wear climate-controlled space suits and live in windbreaks.

The space module is a shirt-sleeve environment driven by fossil fuels controlled collectively. The space suit depends on body heat and relative humidity controlled by the individual. The modern space module's reliability has relegated the traditional space suit to the status of cultural curiosity. With few exceptions, manufacturers design modern space suits to resist only brief periods of exposure to severe outside conditions at high latitudes. Early European expeditions to the Canadian Arctic did not bring dependable space modules. The expeditions that failed to adopt the space suit retreated or perished.

SPACE SUIT PRINCIPLES

The traditional space suit was a powerful tool built on principles that remain valid not just for prolonged-exposure clothing but for design of the space module as well.

Inverted cavity: By trapping warm air on the rise the inverted cavity controls air flow next to the skin. Air in trouser legs warmed by the body cannot rise past the waistband; warm air in the pullover coat cannot rise past the hood. Loosening the waistband or hood releases hot and humid air. Removal of one layer of clothing releases even more.

Still air insulation: Nothing stops the transfer of heat by convection like a vacuum. A space devoid of molecules (which can be excited to a higher energy state) acts as a perfect insulator. Still air is the most common and accessible approximation of a vacuum. Without currents the few molecules present in air are slow to move from one location to another, effectively slowing the transfer of heat. A thin film of still air clings to most surfaces. The two-layered construction of the winter space suit provides five films of still air, one against the body and one on each side of the two layers of fur including the one on the outside (protected from air turbulence by the long guard hairs of the caribou hide). The fur matrix itself, being a still air trap, acts as primary insulator.

Humidity dispersion: The slow circulation of air between the loose-fitting layers prevents humidity caused by perspiration from concentrating at any one spot. Water vapour migrates through the hides and along the hairs toward the outside, where it freezes (to be removed with a snow beater, or left to slowly freeze-dry). The caribou hide "breathes"; vapour does not condense near the body, or soak the hide to diminish its insulation value.

Tailored construction: The suit is made to fit the individual body as to size and personal preferences. No off-the-rack stuff that fits poorly. The tailored space suit has been improved incrementally for more than twenty thousand years.

WINTER SPACE SUIT CONSTRUCTION

Footwear: Up to five layers of caribou or sealskin are used—a stocking with fur facing in; a short boot (hair out); a long boot (hair out); a short overboot (hair out); and next to the sole of the foot a replaceable layer of dried grass, moss, or feathers to absorb shock, improve insulation, and collect excess moisture.

Trousers: Two windproof layers of fur are loosely fitted so that slow air circulation inside, induced by normal body movement, keeps temperature and humidity levels about equal at all points; the fur on the inner layer faces inward; the top is cinched with a drawstring to prevent the escape of warm air.

Coat: Two windproof layers of fur are loosely fitted, with tight-fitting hood to prevent the loss of rising warm air; fur ruffs around the hoods interlock to prevent air loss from the space between the inner and outer layers; the ruffs also reduce wind chill effects on the face (wolverine fur ruffs are preferred, since the slightly tapered hair structure sheds breath condensation ice easily). Warm, humid air not wanted can be expelled by throwing back the hood. Any humidity condensed in the fur at the end of the day has to be removed either by freezing and beating off the resultant ice crystals, or by slow drying inside the snow house.

Mittens: There is one layer of fur (hair out), sometimes with a liner. Since glove fingers expose almost twice as much surface area to the outside as mittens, gloves are never used.

Goggles: These are made of driftwood, horn, or ivory, as protection against snow blindness in spring; horizontal slits reduce light intake without blocking peripheral vision needed to detect lateral movement;

the goggle interior is blackened to reduce internal reflection.

Total weight: five kilograms; very lightweight.

SYMBOLIC CONTENT

Since man and nature are not separate entities in traditional Inuit society, the space suit is both human and animal. Suit designers select fur from the animal's head for the hood; fur from its back for the coat; fur from hindquarters for the trousers; fur from legs for boots and mittens. Patterns of external decoration mimic the animal skeleton. The condition of clothing had meaning: good quality and imaginative decoration show respect for the animals hunted; poor condition signals to the visitor that food is scarce. Inuit culture invests pride in the space suit; European culture invests it in the space module.

SELECTION OF THE FITTEST

The snow house of the Canadian Central Arctic was a sophisticated product of centuries of cultural evolution, a compact bundle of building principles full of lessons for today's designers and builders. Like traditional domestic architecture around the world it is well-worn and honed to perfection—it seems made for its setting. Contrast this to the arbitrary nature of contemporary settlement planning and development.

The snow house fits into its landscape. It nestles in the dense snow drifted into the lee of a rock outcrop; it is made of snow blocks quarried from the same snowdrift. The next blizzard will drift it in further, making it stronger, and warmer to live in. The snow house is custom-built to the size, needs, and whims of the family that occupies it. A group of families makes a semicircular cluster of snow houses spaced about five metres apart; the cluster faces the sea, with entrance ways parallel to the wind so that air currents keep the foreshore clear of snowdrifts.

Demonstration snow house with addition. Repulse Bay—March

Thule house partly reconstructed to show whalebone framing. Resolute—July

The houses may be interconnected by tunnel vaults also made of snow blocks. In the spring the builders leave and the house disappears without a trace—the

Iqaluit—May

*Sod houses.
Uummaannaq—August*

landscape goes quiet again. Intimate in scale and purpose, this architecture suits its builders; in turn it reinforces their pattern of living.

ATTRIBUTES OF THE SNOW HOUSE
Quick to build: It takes one skilled person and a helper about an hour to build a snow house four metres in diameter, large enough to accommodate a family of six. There is quickness in every part of the process. The construction is simple; the main parts are few and interchangeable. The parts are used where they are quarried, reducing transport and double-handling to a minimum. The place quarried becomes the tunnel floor; the snowdrift surface will be the sleeping platform. Once they are laid in a circle, the top of the blocks forming the bottom course is trimmed to form an inclined plane, zero in height at an arbitrary point in the circle to block-height at the end of the full circle (Figure 1-13). The inclined plane guarantees that every block laid down thereafter presses not only downward under the pull of gravity but also against the adjacent block just laid. This prevents each new block, trimmed to lean inward, from falling inward—no time-consuming falsework is needed. Within an hour the loose snow packed into the joints coalesces, converting the block work to a seamless shell. Suspending a sealskin lining from thongs pushed through the wall and toggled on the outside creates an air space against the snow wall. A replacement house, an extra house, a meeting hall cost nearly nothing to build.

Low surface-to-volume ratio: A hemispherical dome uses the least amount of material to enclose a volume above a level floor; it also presents the least surface area to the outside environment, keeping heat loss at a minimum.

High utility-to-weight ratio: A snow block by itself is weak and friable. Snow blocks working together in a hemispherical dome structure have high utility-to-weight ratio. The structure may become strong enough to resist a polar bear's weight—a useful feature in the high arctic. Over time it shrinks inward and settles downward cohesively, without cracking. The structure depends on the shape and thickness of the coalesced snow, augmented by a thin coat of ice on the interior. This glaze is made by closing all openings and heating the empty interior briefly to melt point and then allowing the moist surface to refreeze.

Indigenous materials: The materials used to construct a snow house are local to the site (snow) and to the region (skins).

Weatherproofing: The dome shape offers least resistance to the airflow regardless of wind direction. Wind impact and suction loads that occur remain well below the structure's breaking point. Loose snow (which soon coalesces) piled against the lowest row of blocks counters erosion caused by low-level blowing snow. The wind slowly eats away the snow buffer, leaving the structural snow intact.

Waterproofing: Leaks do not occur in the middle months of winter when precipitation falls as snow. But with the return of drizzle and melt conditions in the spring the snow blocks absorb moisture until the structure can no longer support its own weight.

Thermal resistance: About three quarters of the volume of compact snow is still air, an excellent insulator against the -40°C wind prevailing outside. The inner lining of pelts and the still air behind it allow the inside air temperature to rise above freezing while keeping the median temperature of the wall the few degrees below freezing needed to maintain structural integrity. Drifted snow outside increases the thermal resistance of the dome. With the return of warm days in the spring the snow accumulation outside can be scraped away to reduce the thickness of insulation, and so the median temperature of the wall.

Vapour resistance: Given the frozen-water environment, the confined space, and the human bodies

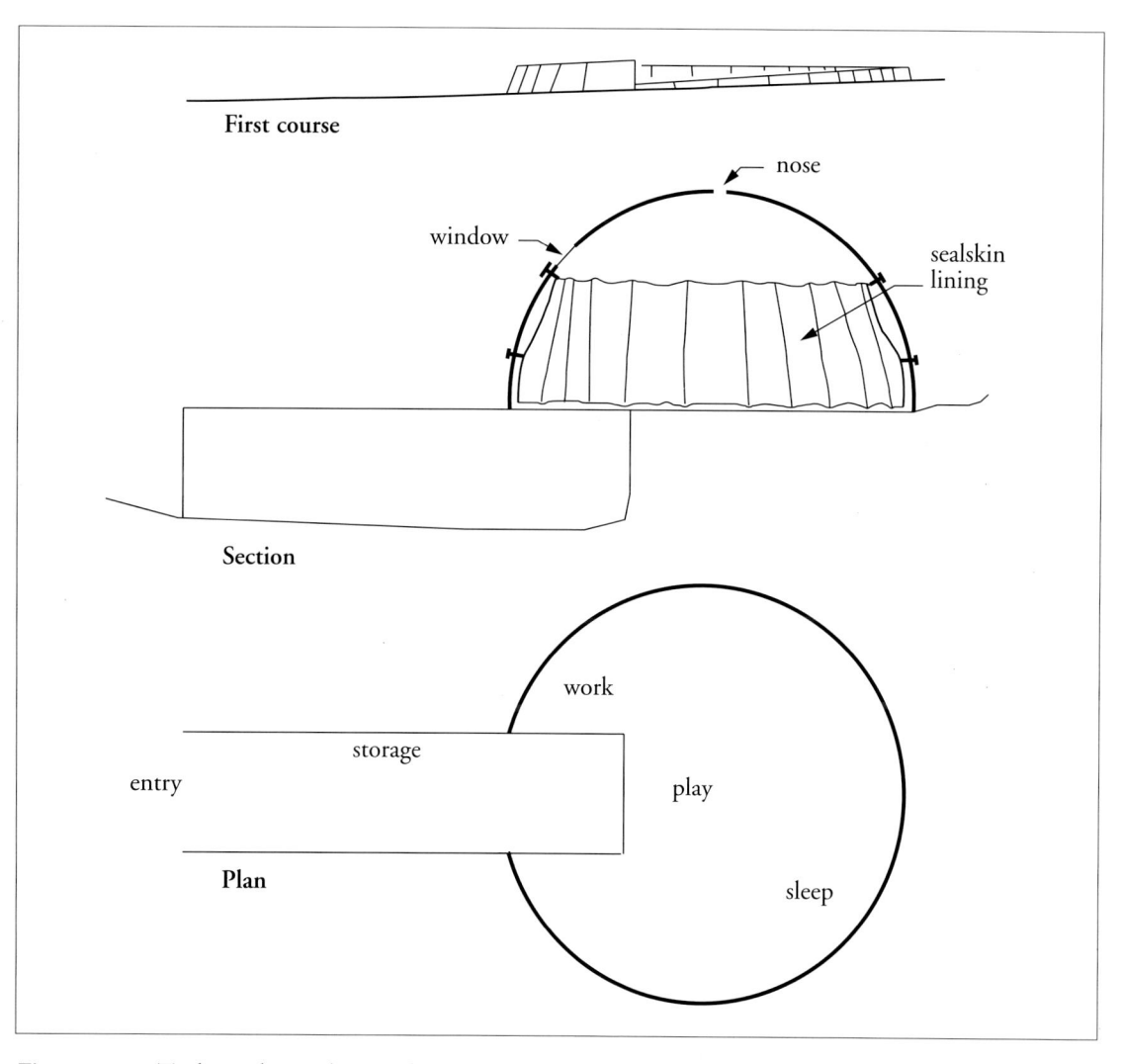

Figure 1-13 Traditional snow house of the Canadian Central Arctic: basic form.

producing vapour, the relative humidity inside the snow house remains high, higher than the relative humidity outside. The seal skin lining acts as a vapour trap by accumulating condensate that refreezes nightly when inside air temperature drops in response to the stillness of the sleeping bodies and the oil lamp. Housekeepers periodically shake or beat the skins to rid them of frozen condensate.

Heat sources: With such good insulation, small surface areas exposed to heat loss, and a comfort level around freezing temperature, the heat source need not be great: body heat and the glow of a seal-oil lamp will do. Nothing else is available.

Inverted-cavity air trap: In winter conserving heat heads the list of priorities. Warm air rises, cold air falls. With no openings in the snow house above the level of the sleeping platform, heated air cannot rise beyond the confines of the hemisphere. Since the entrance tunnel lies about 600 mm below the level of the sleeping platform, cold air cannot enter the inverted cavity as long as the air inside the dome remains trapped. The depressed entrance acts as an air lock. Conversely, as soon as the wad of fur in the "nose" at the ceiling is removed, buoyant warm air rushes out to be replaced by less buoyant cold air from the tunnel below. Leave the vent open too long and the occupants will freeze. Leave it closed too long and the occupants will suffer from oxygen deprivation while the wall temperature rises above freezing. The dome's shape concentrates scarce warm air in the restricted space at and above the sleeping platform—a cube shape would require three times as much warm air to achieve the same level of comfort. A conical dome would leave too much warm air pooled at the apex, well above the sleeping platform. In a still air zone warm air floats above colder air. The air becomes stratified according to temperature. In a snow house this leads to a division of space by activity: housekeepers reserve the warmest air under the ceiling for drying clothes, the middle layer for sitting activity like cooking and sewing, the boundary layer on the platform for sleeping, and the tunnel layer for cold storage.

Flexibility: The snow house does not resist change when new requirements arise; it can be modified quickly. The entrance opening should remain parallel with storm winds to minimize intrusion of snow: a few fresh snow blocks attached to the old as deflectors make an adequate adjustment. A snow block door provides extra protection against storm winds and starving dogs. The tunnel does duty as storage; if more space is required, a small dome can be attached. A line can be fixed to the ceiling to suspend objects needed for fitness games; a tightrope for acrobatics can be run through a large dome from sleds braced against the wall outside acting as anchors.

Daylighting: By night the seal-oil lamp provides a rosy glow; by day blue-green light floods the interior from a sheet ice window in the dome wall above the entrance.

Acoustics: The confined space inside the dome makes speech in low tones practical. Little body heat is lost making oneself heard. The snow wall deadens the sound of the wind; airborne sounds are muted. The snow foundation transmits the sound of footsteps approaching on the snowscape outside; structure-borne sounds are projected.

Drawbacks: This is a damp place. Things and people take a long time to dry out. Respiratory ailments may proliferate.

In today's Canadian Central Arctic the demise of the snow house as a semi-permanent dwelling occurred two generations ago. The snow house had no place in a permanent settlement; few considered it tolerably comfortable compared to a heated wooden building built well. No one wants to go back to living this way.

BAD BUILDINGS

Nothing defeats excellence in architecture as much as low expectations. In the developed world urban populations are so accustomed to buildings that feel and look bad, and may even be dangerous to health, that better is hardly imagined, never expected. So few people have experienced better buildings that a person encountering one for the first time tends to blame an unexpected sensation of well-being on chance rather than on good architecture.

Our basic expectations (shelter, power) exclude anything that makes sense in the larger picture. We find this acceptable in buildings, because humans excel at adapting to new environments no matter how inefficient, uncomfortable, or joyless they may be. Instead of moulding the building to fit our identities, to have it reinforce and improve our ways of life, we allow the building to mould us. We adapt so readily that we can no longer tell whether our acceptance of the new situation is due to our ability to adapt or due to the merits of the building. We try so hard to make the situation work. We have learned to accept that a building costing millions of dollars may be uncomfortable and unresponsive to the real problems of the day, confident that we will somehow "make do." As a society we have lost touch with knowing any better.

Architects must raise expectations by showing through discussion and by example that a building can and should do more than keep out wind and snow: by showing that planned space can and should not just support human activity but enhance it in ways not previously considered possible; by showing that the research performed between the decision to build and the start of design is a vital learning experience for owner, user, and architect with direct consequences for the success of the finished building; by showing that the work of clarifying the purpose and content of the project can be stimulating, enjoyable and, as things start to make complete sense, reassuring; by showing that a successful design evokes in people the agreeable thought that "human intelligence was at work here for our benefit."

Usually this research, variously called programming the building, writing a *program* of requirements, and instructing, or briefing the architect, is done before the owner selects the architect. In a typical project the owner's participation is heaviest at the beginning, declining steadily toward the end as the architect's increases. The owner organizes a site plan, a topographic survey, a soils survey, and a building program with the assistance of surveyors, soils engineers, and building programmers. The owner passes the accumulated data to the architect, who then starts preliminary design work.

INSTRUCTING THE ARCHITECT

The architect expects to learn two things from the building program: what the new building should accomplish, and the physical and social context in which the building is meant to thrive. The purpose may be intuitively clear to the owner and key users—"I know what I like"—but its expression is often inarticulate. Naming a project "elementary school" settles the obvious but says nothing about the quality of the learning to be promoted. There may be disagreement about the meaning of "learning" or about the comparative responsibilities of parents and teachers. Too many people assume that important ideas leap from mind to mind unassisted. Until those ideas surface in a form that can be grasped by each participant, project leaders will conduct their business blindfolded. The quick minds will fall out of step with the slow and the discussion will exhaust itself on details of process, leaving substantive matters to languish in limbo. Key ideas in limbo and bad buildings come from the same seed.

The programming phase was invented to communicate, in intimate detail, the purpose of the building to the architect. Just as important, it serves as a reference against which the architect's performance in design can be measured. Finally, the program imposes project discipline on the owner—he no longer looks to the architect to produce a mindless series of sketch proposals as a means of coming to grips with the purpose of the project.

A good program details type and quality of human activity, numbers, age, and gender of people participating, effective relationships between those people, cultural patterns, salient attributes of spaces, interrelationships of spaces, and relationships of spaces to the whole and to the outside. It follows understanding of the user's needs.

Bridging a cultural gap poses a difficult—some would say impossible—problem for the programmer. Any program depends heavily on unwritten assumptions too numerous and complex to list. If the programmer does not share the cultural background of the user, none of the assumptions made about the user's needs will be reliable. In northern Canada culture gaps exist typically between aboriginal groups and Euro-Canadian settlers. But they also exist between groups of aboriginals, based on differences in language, race, tribal geography, and religious beliefs. Even aboriginals working for the government may find their acceptance in a native community compromised by their supposed allegiance to a system controlled by outsiders.

Consider outside programmers addressing a question thought to be useful and appropriate in their own culture to insiders of another culture. The insiders hesitate, balancing the impact of a question one might consider inappropriate and intrusive with the desire to give an answer that, when applied to a building design, might benefit the community. Or the insiders may believe, consistent with their own culture, that giving an answer calculated to please the outsiders has greater merit than addressing the truth of the matter. Or the outsiders, accustomed to an occasional nod of the head from their listeners to signal continuing attention, may lose patience with insiders who, accustomed to assuming that their presence alone signals paying attention, never nod. Wherever the impact of the measuring stick is greater than the inertia of the object being measured, the measuring stick shifts the object out of position, away from the truth. No measurement—no accurate determination—is possible.

CULTURE GAP
In the middle of the twentieth century, governments pressured the hunter-gatherer societies of northern North America to settle into permanent communities in exchange for government support against epidemic, famine, and "illiteracy." Consider just a handful of the cultural stresses this sea change in lifestyle provoked, and provokes still.

The language of government belonged to the outsider. The message of any meeting was funnelled through interpreters whose inexperience might alter the outcome. Misunderstanding was the rule rather than the exception.

Doing close work on the ground from a squatting position was replaced by working from a chair at a table made unsteady by an uneven floor. The former position exercised all the muscles and joints of the body; sitting in chairs produced back strain, and stiff joints in hips and legs.

Buildings with floors raised above the ground epitomized the cultural divorce from the land. The texture of the land, the smell of the air, and the sound of the birds were baffled by nearly sightless walls made of unfamiliar materials.

The hearth, the fire, or the seal-oil lamp were central to family existence. They threw flickering light and story-telling shadow up to overhead walls that sloped inward like a multi-screened cinema. The oil furnace, with its noise, dust, and stink, replaced the hearth. Firelight's replacement, the incandescent bulb, evaporated shadows or chased them under the new furniture with its overhead, monotonous glare. No substitute—no television set, no living room couch—equals the family-gathering force of the open hearth.

Eating when it is time replaced eating when hungry. Overeating and obesity resulted. The preparer and the consumer of meals had to be at the same spot at the same time day after day.

Sleeping in separate rooms replaced sleeping as an extended family in a single space. Collective selves, forced to spend winter nights as individual selves, developed unfamiliar anxieties.

A house site, the house itself, and their orientation to water, sun, and neighbours used to be a matter of choice. A move to a new location resolved any dislike. Today, a committee armed with a short list of vacant housing, a long list of people waiting for housing, and a point system for ranking the names on the list makes all these choices. Site, house type, orientation, and neighbours never enter the calculation. The committee tells the prospective tenant to take the house or leave it.

CULTURE BRIDGE
Decades of mixed results have caused insiders to view outside visitors to the settlements with distrust. Insiders treat outsiders, particularly those with official good intentions, as less than equals right from the start. Too often outsiders have been taken at their word, taken in and assisted by insiders, only to display a total loss of recall as soon as they leave the settlement. The outsiders fail to complete that most essential aspect of human communication, reporting back to the community (closing the loop) before the next visit on progress arising from agreements made during the previous visit. To insiders the settlement's investment in the visitor shows no profit.

How then do outsiders, trained at great expense on the outside, learn enough about the culture of insiders in the few days and weeks allowed by the project schedule to avoid serious error in the design of a new building for the settlement? In a few words, they do not, because they cannot. Building cultural bridges is easier to say than to demonstrate; building them for a culture in swift transition is more difficult still. In the long term enough insiders, trained on the outside but committed to the inside, will form a pool of talent and wisdom equal to producing architecture that lives and breathes settlement culture. In the short term a rickety bridge across the cultural divide may be better than none. Buildings parachuted into local communities without reference to local conditions do more damage than buildings designed imperfectly with a semblance of cross-cultural understanding.

The programmer aims to elicit cultural information of breadth and depth sufficient to show the designer the way to making a building that reflects the users' past and present way of life. To the user the new space must seem to fit like a glove.

Talk at the start and more talk at intervals during the course of the project is, of course, the main means of eliciting information. To hope to succeed the outsider must leave preconceptions at home. No two remote communities have exactly the same attitude to doing business with the outside; indeed, attitudes change with time and the emergence of new leaders in any human community. Language bars progress instantly. Ideally, the outsider hires an interpreter committed to seeing local cultural values expressed in a clear, dignified way.

Settlement or band councils are the outsider's guide to the community. Once apprised of the outsider's purpose, councillors can identify the most articulate people having information. Organization and interview work properly left to the band or council should not be undertaken by the outsider.

Here are some guidelines to interviewing potential users of buildings: talk not so much to groups, but to individuals, all night if necessary; emphasize

Log walls for the new school. Lutselk'e—July

that the interviewee's opinion, not the interviewer's, will be recorded and applied to design decisions; if unsure of local etiquette, do not apply outsider etiquette, but rather consult a community leader; mirror the body language of the interviewees, for instance, their sitting position, the frequency of their eye contact; speak unassertively, tactfully, slowly, avoiding jargon, allusions, and inside jokes; pay respect by assuming that the interviewees understand; introduce unusual terminology sparingly with the question, "Are you familiar with this term?"; upon hearing an unusual term or phrase ask, "What do you mean when you say ... ?"; where translation is poor look hard for the kernel of a good idea locked in the tortured phrase; ask for examples; allow opportunity to save face; do not hesitate to share personal information.

Go to a community with homework prepared; use questionnaires to acquire a statistical basis for convincing the doubters back on the outside that a community statement has broad support; record or note official meetings; confirm with the attendees that what is recorded matches what has actually been said; keep a diary of meeting and interview dates and locations; do not assess information received on the spot, but rather analyse the total result later; follow fresh leads to the end before departing the settlement.

PROJECT IMPACT

If the blood relationships, cabals, pecking order, motivations, and customs of the settlement seem inscrutable to the outsider, the impact of a major construction project on local society certainly is not. Not so long ago construction companies moved men, materials, and equipment onto the site without notice and without a sideways glance at the local residents, ready to stay for twenty or thirty weeks,

Abandoned housing. Cambridge Bay—March

completely self-sufficient juggernauts. These visitations played out in full view of the whole community. Day after day, residents of all ages and every status faced the fact that total strangers were building a project vital to their development as a society. The money for the building project, perhaps several times the annual budget of the settlement council, was paid out to people who banked every cent outside the settlement. Few benefits of the construction process remained in the community. At present a construction company in the Northwest Territories that fails to show how its presence will benefit the settlement economy risks being left out of the running. Many small communities have since learned to manage the impact of construction projects by demanding—and exercising—more control of the project itself.

Contracting rules can be changed overnight. Social change is more intractable. Construction is the work of tradespeople who have earned their ticket by completing high school education and several years of training and apprenticeship. Workers with this degree of education and experience are scarce in small communities. Training remains beyond the reach of those residents reluctant to leave family, friends, and lifestyle for an outside education that may do no more than make them eligible for the few temporary construction jobs that surface in the community. Permanent construction jobs are few. This vicious circle provides only low-skilled, low-paying jobs to local residents. The slow pace of trades skills development heads the list of problems faced by construction companies operating in northern Canada.

JURISDICTIONS

Construction, since it affects public safety, cannot proceed without official scrutiny. In northern Canada the territorial Fire Marshal rules on safety aspects in buildings by applying the provisions of the National Building Code of Canada. Nevertheless, the onus of ensuring safety in building design remains with the architect. Unfortunately, building codes by nature resist experiment. They reflect current, not future, design practice, construction methods, and materials. In southern Canada the market forces of major population centres can be mobilized by manufacturers to lobby for code changes favourable to new products and methods. In remote areas, particularly at high latitudes, where conditions cry out for fresh approaches to building design, market forces are too slight to impress the editors of building codes. In this sense the code restricts innovation.

Other agencies, such as the Health Department, municipal or regional planning departments, and the Workers Compensation Board also regulate building activity, but given large distances and limited resources the regulatory system depends heavily on self-policing. Institutions founded to regulate the activity of engineers and architects in northern Canada are still in their infancy.

The timing of official inspections of buildings under construction at remote sites is subject to the vagaries of travel and staff availability. Project delays or substandard work often result. Out-of-town inspectors must be kept informed of progress at the construction site in order to minimize the incidence of work delayed for want of an inspection.

PROJECT COSTS

Construction costs in northern Canada exceed those of southern Canada by one third or more. The cost of placing personnel and materials at remote locations accounts for most of this premium. Operating and maintenance costs are also high for the same reason. Legislated trade barriers between territories and provinces increase costs. Since costs cannot be lowered by decree, the reasoning designer examines and compares ratios that improve cost performance.

Comparing construction cost to projected costs of maintenance and operation of the building is essential at any latitude, especially in that future building costs normally exceed today's costs. Reducing the amount of insulation in a building by half may reduce the construction cost by one percent but increase the expense of heating the building over a ten year period by twice the cash value of that one percent.

- Multiple use of space considered at the programming stage may result in lower overall costs. However, multiple use may lead to higher operating costs if special supervision and maintenance become necessary during extended hours of use.
- Every community has a sustainable rate of development. If more development is ordered than the local market can reasonably absorb, prices immediately skyrocket to compensate the contractor for bringing in and accommodating unusually large numbers of scarce tradespersons.
- Selection of compact materials (such as compressible insulation) reduces unit shipping cost.
- Project delays that force shipping and construction into off-seasons cost extra money. Materials that cannot be shipped on open sea lanes must be flown to the site at three times the cost if the project schedule cannot accept a major delay. The productivity of workers exposed to winter weather drops by half compared to the summer level.
- Building designs having twenty different details for similar conditions where five would suffice cost more to build.
- A project requiring seven different types of tradespersons to build costs more than a similar one requiring only five.
- Materials selected from a sole source cost more than materials available from several sources. The absence of price competition for materials supply inflates project cost.
- Methods of construction being tried for the first time cost more than regular methods, since learning-on-the-job slows the pace and reduces the quality of work.

LOGISTICS

TRADITION

Hunting and gathering means following game wherever game travels—the pattern of movement is seasonal but not fixed. One year may see the caribou pass through a major river crossing, the next, nothing. To the hunter, being on the move is normal and being held up in a base camp is unusual. Travel and subsistence are inseparable. Overland travel logistics consist of reducing equipment to essentials (ranging from tent pole and harpoon to thumb-sized amulet) and carrying it on dog sleds in winter or on boats and the backs of humans and dogs in summer. Mobility and acquisitiveness do not mix.

The exception to the keep-only-what-you-can-carry rule is the cache, the place where extra food and equipment can be stored indefinitely—indefinitely, on the understanding that a cache found undisturbed is a gift.

The climate limits mobility. Dogs cannot pull loads in summer or in late winter, when snow and ice go soft. In winter, open leads and pressure ridges restrict travel over sea ice. Spring floods make tributary systems impassable.

PROBLEM

Logistics means putting the specified materials, equipment, and people in the right place on time at reasonable cost. Good building practice at any latitude demands good logistics, but a combination of great distance, limited infrastructure, and uncertain weather makes logistics at high latitudes a special challenge. At lower latitudes a builder has the luxury of rejecting unfit materials or making up shortfalls without delay by calling for new deliveries. At high latitudes there are no alternate sources of supply and no alternate shipping routes to act as safety nets. If the steel-pile foundation is not in place by midsummer, the start of the superstructure will come too late to permit closing-in before the onset of winter. A construction site that is not closed in will fill with snow. Work crews exposed to winter winds will slow to a snail's pace in the darkness: ten months of construction time may be lost.

People commonly ship the wrong quantities of the right material: excess quantity quickly becomes material for unplanned shacks or sleds. Shortage becomes the subject of endless communications between receiver and shipper, each message less clear than the last, describing a situation that no one takes the time to understand fully. Shipments incomplete in scope are not uncommon: plywood but no threaded nails, pipe unions but no joint cement, portable generators but no spare parts. Delay or misapplication of materials results.

For a major structure, perhaps only the foundations and the skeletal frame can be placed during late summer and fall: envelope construction and finishing start the following summer. By naming a completion date for building a project in a remote community and subtracting building season(s), shipping season(s), materials procurement period, and design period, a latest project start date is derived. Missing this critical start date pushes back the completion date of the project one year.

PROCUREMENT

Most building materials destined for use at high latitudes are manufactured at midlatitudes by companies that cater to the local mass market. These companies consider the amount of business done for clients at high latitudes too slight to influence their manufacturing postures. This means that market forces—price competition and reliable service—do not work for the benefit of the high latitude client. Such a client must monitor personal interests closely to be sure that the correct materials arrive at the port of embarkation in good time.

PACKAGING

Since transport has a major impact on the cost and timing of a building project in remote communities it follows that the details of selecting and shipping materials have more than usual significance. A low-volume, light-weight material costs less to ship than a bulkier, heavier one. The two materials may perform equally well in the building. Glass fibre insulation, for instance, can be shipped in highly compressed packages; plastic foam insulation boards cannot. Pre-formed materials such as corrugated steel sheet or acrylic skylights should be nesting—their profile allows them to be packed one inside the other like stackable chairs—in preference to materials that must be packaged and shipped as individual pieces. Light steel Z-girts can be shipped in "nested" bundles; light steel channel studs cannot: they must be bundled individually, greatly increasing the empty volume shipped to site.

Crating for air cargo may be less substantial and heavy than crating required for sea transport, but large crates will not pass through the doors of certain aircraft—larger aircraft may be unavailable, or the destination runway too short to receive larger aircraft. Materials that will not fit into standard containers are also at a disadvantage: they may be shipped as deck cargo—"popcorn"—loaded pell-mell, and exposed to weather and sea.

Material destined for high latitudes must be crated in such a way as to resist damage due to multiple handling: from warehouse at source to truck bed; from truck bed to rail car; from rail car to segregation yard at the rail head; from yard to barge or ship; from barge or ship to beach (sometimes from ship to lighter to beach); from beach to open storage near the building site. Crates do a lot of waiting, usually in the open air. Some materials require weatherproof packaging. Freezing air temperatures destroy certain materials. Frequent wetting and imperfect drying deteriorate others.

LAND TRANSPORT

All-weather highways (western Canadian Arctic) and railways (southwestern Canadian Arctic) make possible year-round transportation for freight and passengers at moderate cost. Building project sites accessible through such systems depend less on season for start and progress of construction. Several of the highways remain bridgeless at major rivers, such as the Liard, the Peel, and the Mackenzie, reducing the all-weather classification by two to three weeks twice a year, at freeze-up and break-up. River ice forms the bridge in winter; motorized ferries make the summer bridge. Projects at remote sites usually depend on the summer sealift, which depends in turn on land transport as a feeder system. In Canada, goods originating at midlatitude may be transported by rail to Churchill, Manitoba, trucked or sent by rail to Hay River, NWT, or to Montreal, Quebec, for transhipment to points further north.

The all-weather highway system can be extended for five to eight weeks every winter by means of a winter road. By midwinter the rivers, lakes, and taiga have frozen deeply enough to support tracked vehicles that clear snow and brush from a route designed to make the least expensive plowing distance between the end of the highway and the off-highway community. Removing the snow cover from the route exposes the ground or lake ice to the colder ambient air, which in turn causes the ground or ice to freeze more deeply, deeply enough to support the high point loads of laden wheeled vehicles for a brief period in late winter. The removal of the snow cover has the opposite effect in early spring: the winter road trace, now exposed both to the higher air temperatures of the new season and the direct rays of the returning sun, begins to thaw in advance of the rest of the landscape. Missing the winter road "window" means delaying the project a full year.

WATER TRANSPORT

Transporting materials in bulk by river and sea remains the cheapest option, despite the restrictions imposed by the climate. For most high latitude destinations the shipping season is both short and inconvenient. It occurs in late summer or early fall, often too late to permit a good start on construction before winter. Pack ice may clog the shipping straits or approaches, preventing scheduled landfalls. Occasionally ice conditions oblige the sealift to bypass a community, severely complicating the local economy for a year. River systems may suffer from low water caused by below-normal precipitation in

Unloading the barge at midnight. Aklavik—July

Unloading the Hercules. Kugluktuk—June

the upstream watershed, or by upstream hydro-electric damming activities. Until the water level rises again, shallow-draft barges must either moor indefinitely in the river or begin the time-consuming business of transferring loads between barges to lighten each barge just enough to clear the shoals.

As soon as the barge or ship arrives, unloading onto the beach begins. Fork-lift trucks move back and forth around the clock, in daylight and in twilight. In the absence of docks, a workable configuration for unloading is developed for each community. Pumps work full time to move fuel from the ship or barge to the tank farm via rigid piping installed above ground from the water's edge.

AIR TRANSPORT

Air cargo is the safety net that saves the bad logistics planners, those people who missed sending their goods by winter road or by sealift. Purchasers of perishable goods, high turnover food items, video equipment, small machines, and machine parts also favour transport by air. For the few communities whose approaches are perennially blocked by pack ice (Pelly Bay) or mountain ranges (Old Crow), transport by air cannot be avoided. Air cargo's speed partially offsets its high cost. In northern Canada air cargo routes are well developed, with the bulk of air cargo moving in turbine or jet aircraft equipped with adjustable passenger/cargo configurations. Jet aircraft are equipped with gravel kits to permit safe movement on gravel runways. Light aircraft can be equipped with skis, wheel-skis, or tundra tires—tires that are oversized, underinflated and lightweight—to suit bumpy terrain where no runway exists.

Scheduled airlines carry most passenger traffic between main points in the Canadian Arctic. Weather delays frustrate both customer and operator, particularly in the central and eastern regions, where fog and blowing snow frequently close airports. Without scheduled passenger service tourism might not exist as an industry in the Canadian Arctic. Airlines provide same-day access to and from most points in southern Canada. Chartered aircraft routinely perform medical evacuations between remote communities and regional centres.

Almost every community has an air strip. Most are too short for large aircraft and some are in poor condition. For large or urgent projects a temporary runway for larger aircraft can be readied on the sea ice.

STORAGE AND ACCOMMODATION

Being expensive to build and operate at high latitudes, heated storage space is scarce in remote communities. A small amount of covered cold storage may exist, frequently consisting of older buildings on the verge of abandonment. The latter may leak or fill up with powder snow that melts in the spring. This leaves open tundra or clear-cut taiga for common storage. Many building materials can survive this kind of exposure for several months if properly packaged. They may not survive pilferage so readily. Crates left in the open will disappear under snowdrifts wherever blowing snow occurs.

Temporary accommodation in small communities varies from the gymnasium floor to modest hotels equipped with eight bedrooms, two or more persons to a room.

EQUIPMENT

Renting standard construction equipment in small remote communities at high latitudes in Canada may not be possible. For instance, there are no mobile cranes, apart from the occasional flat bed truck with a lifting arm, or the power company's cherry picker. Even if rental of existing equipment can be negotiated for a reasonable price, the actual working condition of said equipment on the day the building contractor needs it cannot be predicted. The local authority controls most heavy equipment of use during a building project—regular duties such as snow removal, runway maintenance, and gravel haul may limit availability of equipment to outsiders.

Pile-driving equipment located in one community can be transported to another that needs it only with careful planning, since the shipping season limits such movements to one per year per machine. High costs and short runways may deter selection of the airlift option. Advantage can be taken of the occasional visits of special equipment to a given community—rock crushers to make gravel make an appearance when a runway must be extended or improved, so stockpiling gravel for future unrelated projects becomes a possibility. Compaction equipment is as scarce as rock crushers. There may even be no batch mixer for concrete on site.

COMMUNICATION

Good lines of communication are vital to a building project. A century ago an exchange of letters at high latitudes in Canada might have taken several years. In the first half of the twentieth century the turnaround was reduced to a year. Now most remote settlements possess at least one dish antenna pointed at a satellite in geostationary orbit, making nearly instantaneous exchange of voice, video, and data an unremarkable phenomenon. Private telex service exists at a few major institutions and businesses. The telefacsimile machine has made huge inroads since its introduction. Besides its ability to transmit and receive images and text at any time of day, the fax can be operated by unskilled persons—the communications specialist has been eliminated. Barely literate persons with a good idea to communicate no longer hesitate to give the receiver a piece of their minds. Electronic mail networks, while growing as a medium, so far do not offer simple access.

Community radio dominates the airwaves in most remote settlements. Some television programming originates in regional centres, but most comes from midlatitude sources broadcasting around the clock. Most households possess a video cassette recorder. Cellular phone nets exist in major centres. Citizen band radio for local communication between house or office and vehicles in town or on the land complements the telephone system. In the western Canadian Arctic a radio-telephone system exists to link households and vehicles scattered throughout the Liard-Slave-Mackenzie basin.

Dish aimed at a geostationary satellite. Upernavik—August

BARRIERS 16

TRADITION

People living on the land developed space suits as primary thermal, air, and rain barriers. Their tents and snow houses worked as back-up, not substitute, barriers. There were no vapour barriers. Water vapour moved around humans with little interference. Damp clothing had to be taken out of service and dried out near the fire.

PROBLEM

Just sixty years ago central heating and insulated walls were a novelty in North America. The expression "building envelope" had no meaning. With increases in the standard of living, the pressure to improve comfort by building better envelopes became relentless. Once freed of wartime priorities, industry turned to developing the materials needed to satisfy the demand for built-in comfort. Product research was brief—it was trial-and-error—and the building owner absorbed the cost of product failure in a building.

The failure to realize that products slapped together in a building envelope work together, or against each other, twenty-four hours a day compounded the problem. Each product modified the effectiveness of all the other products that made up the building envelope. Sixty years ago little was understood about these side effects. A pre-World War II wood frame house, for example, consisted of clapboard over wood studs on the exterior and lath and plaster on the interior, with nothing in between but mouse warrens. The walls were drafty. In winter the hot air rising in the chimney above the fireplace sucked in cold outside air through the floors and lower walls. Heated air exited through hairline cracks high in the walls and around windows. The cold outside air, being low in moisture content, became relatively drier when heated. Still, some of the water vapour inside, for lack of an air or vapour barrier, passed into the envelope assembly itself. The risk of rot remained low, since the mean temperature of the envelope in winter was not high enough to support decomposition, and the vapour that did condense there was soon re-evaporated by the drafts penetrating the clapboard.

The simple act of stuffing insulation into the concealed cavities of the wood frame resulted in higher mean temperatures for the envelope. The presence of insulation also reduced the velocity of air currents, so less condensed vapour re-evaporated. More water and more heat inside the envelope assembly promoted more rot in structural wood. So the improvement in thermal comfort for occupants—adding insulation—led to deterioration in structure. Similar improvements made with little regard for side effects abound in the history of building envelope design.

Many agencies now study and test building products and prototype envelopes. Nevertheless, the avalanche of new building products reaching the market every year far outstrips the existing laboratories' capacity to test them all, even superficially. As a result, a manufacturer can still turn a profit by convincing owners to wrap their houses with a new product, even though the benefits of doing so may not be striking, and the product may cause no visible damage to the envelope for the first fifteen years.

Finally, public confusion about building envelope components and their purposes persists. The building scientist uses precise terminology unfamiliar to the builder, while the scientist has little of the day-to-day experience of construction that is the builder's bread and butter. The builder has no scientific background, so that the profound reasons for doing things in a certain way do not come to mind—the builder does things today the same way they have been done for the last five years, right or wrong. The builder prefers not to spend time thinking through a problem that, more or less successfully, has been solved by trial and error. Architects and engineers must work in both worlds, and as a consequence know a little about everything but not a lot about the specific problem at hand. Owners, on the other hand, hire architects and engineers to think things through, to reduce the general confusion.

CONCEPTS

There is no such thing as a perfect barrier in a building envelope. Trace amounts of the gas or liquid being excluded always get through, and to other gases and liquids the barrier may not be a barrier at all. A thin film of plastic used as a vapour barrier blocks most but not all of the water vapour pressing against it; consequently, it only retards the flow of vapour. Given time the high moisture content on one side of a film stretched across a sealed box will eventually drop to match the ambient moisture content on the other side. But if it is unsupported against air currents, the plastic film bellies out and eventually tears to shreds. A six millimetre thick sheet of stainless steel would resist both air and water vapour better—but at what cost? The word "barrier" is simply a convention used to differentiate things as apparently substantial as a blank wall from sieves, screens, and filters. In truth, a blank wall is little more than a fine-meshed screen that paint particles cannot fall through.

To act as a barrier against liquids and gases a material must be continuous, durable, and stiff enough to resist anticipated pressures. A roofing material that loses flexibility when exposed to sunlight is no good; an insulation that loses thermal resistance by slumping in a cavity under the pull of gravity is no good; an air barrier material that is cut into small, "manageable" pieces and installed with unsealed laps is no good.

Keeping track of the trace amounts of the gas or liquid that manage to penetrate the barrier equals in importance keeping out the significant amounts. Small amounts of trouble can do large amounts of damage given sufficient time. Trace amounts, once past the barrier, should be permitted to escape the envelope readily. A second barrier to contain them would serve only to trap the trouble inside the system indefinitely, precisely where it can do the most damage. Place a good quality barrier in the best location possible within an envelope assembly and then see to it that anything that passes through for any reason can escape to the outside. Two barriers of the same type are not better than one.

Backup systems keep astronauts and air travellers among the living. High latitude buildings must have backup systems as well. For instance, since all roofs leak eventually, a roof with a vapour barrier designed to conduct roof leak water to the edge of the building where it can drain to the outside is superior to one whose vapour barrier performs like a sieve under a kitchen tap. A vapour barrier can be designed to act as backup for the rain barrier. The designer's solution must reconcile the apparent contradiction between "backups are good" and "one barrier is better than two."

THERMAL BARRIER

Insulation does not stop the flow of heat from inside to outside, it just slows it down. The thicker the insulation, the slower the outward heat flow, the lesser the amount of heat that has to be generated inside the building to compensate for the loss. But there is a practical limit. Insulation is expensive; so is the shipping and installation of it. Excess insulation will cost more than the expense of the extra heating fuel needed if the excess insulation did not exist. The happy medium can be calculated from the cost of insulation delivered to site, installation costs, annual degree-days, outside design temperature at the site, and the current and estimated future price of heating fuel. At high latitudes the result, stated in thermal resistance units, usually approximates Government of Canada guidelines for housing in geographic zones having annual degree-day totals in excess of 8,000: RSI 2.2 for foundation walls, RSI 4.7 for floors, RSI 4.5 for walls, and RSI 7.1 for roofs. The guidelines include separate, lower figures for large buildings.

A wall containing RSI 3.5 insulation is usually said to have a total thermal resistance of RSI 3.5. This is misleading. It is particularly misleading for envelopes in which structural members or openings interrupt the continuity of the insulation. Heat flows through a wood stud three times as rapidly as it does

Exterior walls under construction. Rankin Inlet—October

Heated swimming pool at latitude 76° N. Little Cornwallis Island—July

through the same thickness of insulation. Compared to the insulation, the wood stud acts as a *thermal bridge* between the warm inside and the cold outside. Calculation of an envelope's average thermal resistance must therefore factor in the total area of "cold spots."

Restricting the extent of thermal bridges is important not just to reduce total heat loss but also to avoid the condensation that arises when warm inside air meets a cold surface. A material that conducts heat quickly to the outside will be cold on the inside. Thermal bridge patterns show up clearly on infra-red sensitive camera films. On a windless day following a blizzard, residual patches of snow sticking to an exterior wall mimic the pattern of the studs concealed by the wall cladding. Where the exterior face of the wall is coldest (well insulated) the snow sticks for hours or days. Where the wall is warmest (at poorly insulated stud locations) the snow sublimates quickly, exposing the colour of the wall.

Insulating materials lose their thermal resistance over time. Glass batt insulations tend to settle, losing loft or thickness; plastic foam insulations tend to lose the gases used to form the bubble matrix during manufacture. This gradual loss of thermal resistance increases heating costs as the building gets older. Thermal resistance also decreases when the design of the envelope permits air to move too freely through the insulation. The combination of stack effect and air leaks in the envelope means that cold air entering at the bottom of the envelope replaces warm air lost at the top. Simple convection patterns inside insulated cavities also have a detrimental effect: air inside the wall rises on contact with warm inside surfaces only to cool and descend again on contact with the cold outside surfaces. This cycle speeds the flow of heat to the outside. Restraining vertical air flow inside the envelope minimizes heat loss. Insulation materials contaminated by water or snow also lose thermal resistance—water conducts heat more rapidly than plastic foam or glass fibre.

Insulation demands a relatively calm, moisture-free environment to work effectively. An envelope having just a barrier, without barriers against air, vapour, and rain, will have poor thermal resistance.

AIR BARRIER

In recent decades designers and builders believed that the vapour barrier was the key to the modern insulated wall. A sheet of plastic nearly impervious to water vapour placed on the warm side of the envelope would prevent the migration of water vapour by molecular diffusion from the enclosed space to the cooler insulated cavities. Previous experience had made it clear that water vapour condensing inside an insulated wall promotes rot in wood structures and contaminates the insulation as well. But in many new buildings the addition of a vapour barrier did not eliminate wood rot and contamination of insulation. The reopening and inspection of many such failing envelopes led to the conclusion that the water vapour problem was due mainly to the migration of water vapour with air currents, rather than migration by molecular diffusion between zones of different moisture content. Even quite small holes in the building envelope permitted large amounts of air, propelled by stack effect or wind forces, to pass through over long periods of time, say a winter season. Air currents became the chief villain, the main cause of condensation inside the envelope assembly.

About this time it also became clear that plastic film used as a vapour barrier, supported only by a few staples, was no match for the countless air pressure changes exerted on buildings by the wind and by stack effect over extended periods of time. The plastic simply tore away from supporting staples, the laps opened up, and the material itself began to fatigue under the constant flexing. All this happened inside the envelope assembly, out of sight and out of mind, beyond the reach of casual inspection.

Although designing structural members of buildings to resist wind loads was routine, the idea of preventing the passage of air and vapour through the envelope presented a new challenge. Like a vapour barrier, an air barrier demanded continuity throughout and in detail; unlike the vapour barrier it had to resist wind impact and suction forces. On one side of a building the wind pushes; on the other it pulls. As soon as the wind direction changes the forces acting on the envelope may be reversed.

An air barrier material, no matter how impervious to air molecules, fails to act as an air barrier system in a building unless it is both continuous and fully supported against wind impact and suction forces. A thin membrane marketed as an air barrier fails to act as an air barrier system unless sealed at all joints and supported continuously on both sides. A concrete block wall fails to act as an air barrier system, even if properly supported, because shrinkage cracks open in the mortar and structural cracks open in the blocks. Air passes through concrete block as though through a sieve. Rigid sheets of plywood fail to act as an air barrier system unless supported at suitable intervals and at all joints. Gypsum board cannot act as an air barrier system unless cracking due to shrinkage and foundation movement can be avoided.

Any air barrier system placed in a wall at some distance from the plane of the vapour barrier must itself be vapour-permeable to avoid trapping water vapour between the two barriers.

Exterior walls and roof under construction at Detah. Yellowknife—July

VAPOUR BARRIER

At high latitudes in winter the moisture content of air inside a heated building is usually higher than the moisture content of the outside air. Water vapour molecules migrate across a barrier from air with high moisture content to air with low moisture content. Water vapour molecules diffusing through the building envelope condense inside the envelope assembly on contact with colder air and colder surfaces. This condensation may be the last straw—the last of the conditions necessary for rot in wood to start. Condensation also reduces the efficiency of the insulation present. Since water vapour has a greater heat capacity than dry air, the loss of the heat stored in water vapour escaping through the envelope becomes significant over time as well.

Water vapour migrates by molecular diffusion, a flow of water molecules from a parcel of air crowded with water molecules to a parcel of air not so crowded.

A cloud of water vapour, although invisible, expands, thins, and disperses like a cloud of smoke. Erecting a barrier with holes small enough to stop most water vapour molecules will impede this migration. A thin plastic film, a thin film of aluminum, two coats of oil-based paint, or a sheet of water-proof plywood all have published quality ratings as vapour barriers. Clapboard, stucco, bricks, and roofing shingles do not rate as vapour barriers.

RAIN BARRIER

The rain barrier exists to prevent rain and snow from wetting the envelope assembly and the structure. Wherever leaks in the air barrier system occur, differential air pressures will draw rain and snow past the rain barrier. Fine snow may penetrate the envelope altogether and form a drift on the floor inside, or it may simply nest undetected inside the envelope, where it remains frozen until spring.

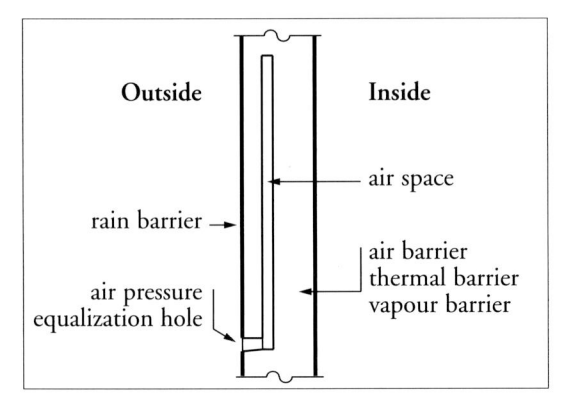

Figure 1-16 Principal components of the rain screen shown schematically in a wall section.

To prevent rain and snow migration across the rain barrier due to differential air pressures—high on the outer face, low on the inner face—an air barrier system is required. Placed somewhere inward of the rain barrier (so that it does not get soaked by rain), the air barrier system resists wind pressure. This resistance, coupled with small openings placed at intervals in the rain barrier, allows instantaneous air-pressure equalization across the rain barrier. That is, air pressure on both sides stays in equilibrium no matter how gusty the wind. In brickwork the openings are formed by leaving a few vertical mortar joints open, unmortared; in wood siding the standard joints between boards are sufficiently leaky to permit air-pressure equalization. With the air pressure equal on both sides of the rain barrier, no force exists to drive rain or snow horizontally beyond the rain barrier. No force, that is, except capillarity. To prevent water absorbed by the rain barrier material from "wicking" over to the rest of the envelope assembly a vertical air gap (or compartment) is left. The air gap also speeds up the re-evaporation of any water absorbed by the rain barrier. A rain barrier with small openings at intervals and an air barrier system working behind it are together called a rain screen. Figure 1-16 shows the rain screen principle schematically.

The expansive forces of the freeze-thaw cycle deteriorate most building materials. Wood absorbs the punishment for a long time; not so the paint or sealer that covers it. Brickwork that freezes after absorbing large quantities of moisture will spall, exposing even more absorptive surfaces to water. Resistance to the corrosive action of water mixed with local and atmospheric constituents and resistance to sunlight also affect the selection of rain barrier materials and finishes. The rain barrier must be vapour-permeable to avoid creating a vapour trap outside the vapour barrier, unless other steps are taken to disperse vapour concentrations.

BARRIER LOCATIONS

An envelope that separates a heated interior from high latitude winter weather requires the presence of four barriers: thermal, air, vapour, and rain. From the inside out there is an interior finish layer, a vapour barrier, a thermal barrier, and an air barrier, followed by an air gap, and a rain barrier. The rain barrier deflects most of the incident rain away from the building, first absorbing and then re-evaporating the rest. The air gap prevents water drawn by capillary action through the rain barrier from reaching the air barrier. The air barrier prevents the incursion of airborne rain and snow beyond the rain barrier; it also prevents the passage of air currents into and out of the building, and in this location reduces local circulation of air currents between thermal barriers made of batt-type insulation and the outside. In this position the air barrier material must be vapour permeable—any water vapour escaping outward through the vapour barrier can pass to the outside. The thermal barrier reduces heat flow, without resisting the passage of vapour toward the outside. The vapour barrier slows the molecular diffusion of vapour into the envelope assembly.

V
DESIGN

17 PLACE
TRADITION *88*
PROBLEM *88*
STRATEGIES *90*
SETTLEMENT SITING CRITERIA *90*

18 NEIGHBOURHOOD
TRADITION *92*
COMMUNITY *92*
STARTING SMALL *93*
ORIENTATION *94*
SNOWDRIFTING *95*
PRIMACY OF THE VEHICLE *97*

19 GEOMETRY
TRADITION *98*
PROBLEM *98*
PROPORTION *98*
PERIMETER-TO-AREA RATIO *99*
SURFACE-TO-VOLUME RATIO *100*
CONCENTRIC ENVELOPES *101*
JOINTS *101*
LAPS *103*
SEALANT SHAPE *103*
INVERSE SQUARE LAW *103*
SCALE MODELS *104*

20 PLANNING
TRADITION *105*
PROBLEM *105*
CIRCULATION *105*
SCALE *106*
BOUNDARIES *106*
STRUCTURE *106*
FLOOR *107*
STORAGE *108*

21 ENTRIES
TRADITION *109*
PROBLEM *109*
STRATEGIES *109*

22 FOUNDATIONS
TRADITION *113*
PROBLEM *113*
CONCEPTS *115*
SITE SELECTION *118*
SITE INVESTIGATION *118*
STRATEGIES *119*
FOUNDATION TYPES *119*
PITFALLS *119*
PERFORMANCE *121*

23 WALLS
TRADITION *122*
PROBLEM *122*
STRATEGIES *122*
WALL ASSEMBLY TYPES *124*
EXAMPLE *124*

From one extreme...
Gjoa Haven—February

...to the other. Igloolik—April

24 WINDOWS
TRADITION *126*
PROBLEM *126*
STRATEGIES *126*
BARRIERS *128*
WINDOW TYPES *129*
SKYLIGHTS *129*
DAYLIGHT *129*
FRAME *130*
MATERIALS *130*
HARDWARE *131*
OTHER FEATURES *131*

25 DOORS
TRADITION *132*
PROBLEM *132*
STRATEGIES *132*
BARRIERS *133*
DOOR TYPES *133*
HARDWARE *133*
TELLTALE *134*

26 ROOFS
TRADITION *135*
PROBLEM *135*
CONDITIONS *136*
STRATEGIES *137*
ROOF ASSEMBLY TYPES *139*
ROOF INSULATION *139*

27 MATERIALS
TRADITION *141*
INDIGENOUS MATERIALS *141*
MODERN MATERIALS *142*
PERFORMANCE *142*
DEGRADATION *142*
PREFABRICATION *143*
CONCRETE *143*
MASONRY *143*
METALS *144*
WOOD *144*
PLASTICS *145*
GYPSUM BOARD *145*
COATINGS *145*
COLOUR AND TEXTURE *146*

28 SERVICES
TRADITION *147*
IN THE FIRST PLACE *147*
HEAT GAIN, HEAT LOSS *147*
STACK EFFECT *148*
VENTILATION *149*
HEATING *149*
WATER *150*
SEWAGE *150*
SPRINKLERS *151*
FUELS *151*
UTILIDORS *151*
ELECTRICITY *151*
LIGHTING *152*
GARBAGE DISPOSAL *152*
SERVICE ENTRY CO-ORDINATION *152*
INTERIOR CO-ORDINATION *153*
MAINTENANCE *153*

4-unit cluster housing for independent senior citizens. Yellowknife—April

Student residence bedroom. Rankin Inlet

PLACE

Ancestral campsite with access to the water margin. Devon Island—August

TRADITION

To high latitude societies dependent on the hunt, being in the right place at the right time of year was crucial in the choice of place to live. They made the best of the places where, for a few days or weeks, the fish usually ran or the caribou usually crossed. Permanence meant returning to the same places every year, for all or part of the season, as long as game was plentiful. Permanence today means settling in a single place. The experience needed to select a place for permanent, year-round settlement—a place for all seasons—never developed.

People on the land needed to know how to site a temporary camp. Siting a tent in haste only to have the prevailing wind gather inside and explode it teaches the traveller precisely where not to pitch the tent next season. Personal experience, coupled with the experience of elders, makes proper siting second nature.

Still more important, people needed to know how to site a seasonal camp. Skraeling Island, the largest of several small islands in Alexandra Fiord, on eastern Ellesmere Island, contains the ruins of houses and meat caches abandoned by Thule people about 800 years ago. Artifacts excavated there and elsewhere in the Arctic in the last fifty years suggest that these people subsisted primarily by harvesting marine mammals. The presence of several polynyas in the vicinity of Alexandra Fiord attracted populations of walrus, whale, and seal. Both major housing sites on Skraeling Island exhibit similar evidence of deliberate site selection. Several metres above high water mark, they occupy gently sloping ground consisting of grassy sod, which provides depth for excavation and material for building; they have easy access to the sea via narrow, gravel beaches for launching boats and beaching catches; they afford simultaneous views of sea mammal activity on the surfaces of Buchanan Bay (to the east) and Alexandra Fiord proper (to the west); they face the sun to the south; they lie adjacent to rock outcrops that shield the houses from northerly winds.

At present a small group of Slavey Dene occupy a site every spring on the north bank of the Loche River next to Willow Lake, 40 kilometres north of Tulita in the western Northwest Territories. The Willow Lakers go there to take fish, muskrat, and rabbit. Five or six families camp in an assortment of log cabins, wall tents, and tipis strung out east-west along the river bank. The campsite faces south, in the direction of warm spring sunlight, and in the direction of the river, the main means of travel to and from the camp. The camp site backs onto northerly winds and an arm of Willow Lake. The smoke from wood fires drifts down to the riverbank.

PROBLEM

Deciding where to place a new permanent community, or even a new neighbourhood in an old community, outweighs other decisions affecting the vitality of the settlement. People with enough experience and confidence to make the right decision the first

Campsite. Kitikmeot Region—July

Village plan seen from the air. Inukjuak—April

time seem to be lacking at present. Into the vacuum tumble all sorts of considerations, the real and the imagined given equal or arbitrary weight. Sorting goes by discussion and disinformation; a plan congeals. Some oddities result from this process.

The community of Rankin Inlet coalesced around the pithead of a new mine. Although the mine closed five years later, the settlement found new reasons for being. It soldiers on, its location decided by mine-working criteria that have vanished from memory.

Resolute Bay began at the whim of the 1947 summer ice pack. Extending farther south and east than usual, the ice turned back the ship bringing prefabricated weather stations to the Canadian high arctic for the first time. Since this put the planned destinations beyond reach, the expedition was over before it began. The ship was ordered back to Boston. To salvage something useful from the mission its leaders lobbied for government approval to offload the expedition at the nearest available beachhead. When the order to disembark finally came over the radio, Resolute Bay (an uninhabited inlet of Cornwallis Island) happened to be abeam the returning ship. In terms of safety and convenience for ships and aircraft a worse site could not have been chosen, yet Resolute became the transportation hub of the high arctic archipelago. The enormous size of the investment in Resolute—extra-long runway, hangars, housing, and equipment—nipped in the bud any proposals to move the settlement to a better site.

Near the peak of regional oil and gas exploration in 1974 the territorial government did select a new site for the residential section of Resolute. Government consultants evaluated six potential sites, all within a few kilometres of Resolute Bay airport, a base of activity still too important to abandon. Through town meetings the planners consulted Resolute's residents on site selection. The residents seemed to favour a site also favoured by the planners. Later it was discovered that they favoured the other sites just as much, unwittingly making a mockery of the consultations. Construction of New Resolute began but stalled almost immediately as the oil and gas bubble burst for lack of cheap, reliable transport to southern markets. Instead of building a new town the government's program degenerated to one of relocating the residential buildings from the old site to the new one.

DESIGN

Modern town with access to the water margin. Uummannaq—August

The Resolute New Town exercise did point up two basic problems: first, the meaning of consultation differs among different cultures and, second, local residents new to living in permanent settlements do not necessarily understand more about site selection for permanent high latitude settlements than any other interested party.

Judging from the locations of existing high latitude communities in Canada one requirement for siting remains uncontested: proximity to the water's edge. At least one edge of the settlement must reach the sea, or the river, or the lake. No planning committee can muddy the symbolic and practical importance of the water margin.

STRATEGIES

By squandering the advantages offered by the terrain a poor site robs a settlement of vitality. A poor site costs more to operate and maintain than a good one. It is important to know, understand, and exploit the settlement's reason for being; to consult the users before proposing planning principles; to devise a program of requirements for approval by consensus. Arbitrary decisions have short lives but long-term consequences. The recurring fundamentals are:

1. Face the sun at the equator.
2. Face away from the polar wind.
3. Locate next to the water.
4. Choose a slope that drains meltwater and cold air.
5. Avoid depressions in terrain that accumulate blowing snow.
6. Find solid ground, preferably in continuous permafrost, or in soil without permafrost.
7. Select a place with room for expansion.

SETTLEMENT SITING CRITERIA

Sun: Slopes facing the equator absorb more solar radiation and shed snow cover earlier in spring than others. Slopes facing the pole remain under snow well into summer, sometimes perennially. A place shrouded in cloud or fog much of the year (such as Resolute, which occasionally goes without a summer) has little to offer. A site with a mountain between it and the equator is not much better. High latitude sunlight, low-angled much of the time, throws long shadows across the landscape even from small natural obstructions, delaying snow melt. Select a site that allows penetration of low-angle sunlight from almost any point on the horizon, as a means of banning zones of permanent shade.

Wind: Select a place whose features diminish the impact of polar winds. Avoid any site at or near the low end of a topographic funnel, to reduce the effects of chinook winds descending from the mountains. To minimize the danger of advancing fire fronts within the tree line choose a site, with natural firebreaks—exposed ridges, wide waterways—located in the forest upwind of the community.

Water margin: Whether needed as a commercial dock for the fishery, for loading and offloading barges and ships, for beaching whales or motorboats or float planes, for swimming or skating, for watching the start and return of the hunt, the water margin represents the settlement's prime asset. Select it carefully. See how it behaves in all seasons. Do summer winds blockade the landing with pack ice? How does the tide affect it? Is the bottom suitable for stranding ships? Are there safe moorings? How does spring breakup affect it? Is there flooding? Is there potential for erosion? Can it be developed as a harbour later? A seasonal waterway is essential, but it takes a road connection between the settlement site and larger centres to keep the retail price of everything from oranges to windows in check. Every community requires a reliable supply of potable water, preferably a natural lake.

Drainage: Find a site on a low slope that allows the coldest air to drain to lower elevations. Since water will not percolate through perennially frozen soil, it must drain laterally, rise in vegetation, or evaporate. Avoid sites with poor drainage and standing water.

Snowdrifting: Blowing snow left unchecked will settle in windward neighbourhoods and buildings, potentially raising the surface of the snow to eaves level. Beyond the tree line, flat but rough-textured (pockmarked) ground upwind of the settlement site will serve as safe storage for drifting snow. Within the tree line a stand of conifers a good distance upwind of the community will have a similar effect. The amount of snow available to be picked up by wind each winter is finite—if the blowing snow headed for the settlement can be made to drop out of the airstream at an upwind location, many of the problems associated with drifting snow will not arise. An absolutely flat upwind topography stores little snow and allows the wind to accelerate before reaching the settlement. Inevitably some snowdrifting will occur, creating removal and storage problems. Storing snow in artificial mountains on flat sites merely compounds accumulation problems during the next blowing snow episode. A sloped site, particularly one with back slopes here and there, provides snow storage locations conveniently close to accumulation points. In spring, large snowdrifts may behave like dams about to burst—meltwater must be allowed to drain quickly or damage to roads and foundations due to erosion will occur. A site in the mountains must be checked for avalanche potential.

Soil: Exposed bedrock and raised beaches usually offer stable foundation prospects. Very fine-grained soils are frost-susceptible—they tend to attract and store free water—and frequently lead to foundation failure. Ice-rich soil does not exhibit any behaviour pattern compatible with foundations for buildings. Bogs and sedge meadows are breeding grounds for mosquitoes. Ice-rich soil over sloped, permanently frozen ground will creep a few centimetres downhill every summer—*solifluction.* Placing foundations in ground with discontinuous permafrost demands special attention, partly because one local area may be stable over permanently frozen ground while another settles slowly over unfrozen ground, and partly because discontinuous permafrost, being so close to thawing temperature, is volatile—the smallest microclimate changes may cause it to disappear or expand in the course of a single season.

In soils with continuous permafrost, water near the surface cannot always be avoided. Water purity may be high in remote areas, so high in fact that soaked soil performs poorly as grounding for electrical systems, another criterion for careful siting.

Expansion: Here as anywhere else in the world a new community must have a reserve of land suitable for later development at reasonable cost. As soon as all the suitable land is absorbed by a growing community the cost of land will soar.

Other site resources: Modern high latitude settlements cannot survive without accessible deposits of natural gravel for constructing roads, airstrips, and foundations for buildings. Every remote community needs a nearby site suitable for a 600- to 900-metre-long airstrip complete with glide paths free of obstructions. An adjacent frozen waterway with ice characteristics suitable for reinforcement to permit landing of large air freighters may substitute for a long runway on land. Stable, leakproof areas must be found to serve as sewage lagoons and waste disposal sites. Tank farms for storing fuel have similar requirements—stable foundations for tanks, emergency fuel containment in case of tank failure, and proximity to barge moorings for easy fuel transfer. Cemetery ground must be located. Wildlife zones (for example, snow goose nesting areas, polar bear routes) require acknowledgment and protection.

NEIGHBOURHOOD 18

Rock contours sway the town plan. Uummannaq—August

TRADITION

For the parents and grandparents of the last two generations there is a certain sense of exasperation every spring. Like sled dogs impatient for autumn's first snowfall, they strain at the bonds preventing the move from the permanent settlement to the campsite on the land of their forebears, land just renewed by frost, snow melt, and waxing sunlight. Anxious to quit their hot, stuffy box of a house, they must be patient a while longer, stymied by school hours and school days that, however promising for the children and grandchildren, seem to chain life to a jerky rhythm completely out of step with the land.

For the first time in a year, separate households bursting with progeny can live next to each other in tents grouped at the water's edge precisely in accordance with people's preferences. Kinship and friendship govern the layout of this neighbourhood. Not subdivisions set out to minimize costs of pipe and cable. Not distant bureaucracies, procedural rules, and outside planners. Not house designs with entries facing upwind and views boxed in. Just tent walls pervious to the sound of ducks, loons, arriving boats, distant conversation, and children playing past midnight. Tent walls pervious to the light of the sun, warming sleeping bodies into wakefulness, and to the shadows of passers-by. Tent walls pervious to the smells of fresh water, new moss, moist ground, and the camp fire mixture of willow smoke and fried whitefish. Tent walls pervious to the light breeze that keeps mosquitoes at bay and the structure in motion, mimicking life itself. Tent walls pervious to the decision to move from the site, almost without leaving a trace, like life itself.

COMMUNITY

If the old community spirit still exists in the permanent settlements it is no thanks to the built environment. Community spirit must often survive in spite of settlement design. In Canada, the high latitude community that properly serves the needs and aspirations of its population has not been built.

Subdivision grid patterns originally devised to unroll suburban fabric over the flat agricultural land surrounding southern Canadian cities now appear on the tundra and in the boreal forest. The grid balances two opposing tendencies, the need to make the pattern as small as possible to reduce costs of development and services, and the need to separate buildings by at least twelve metres to control the spread of fire. The unfolding pattern takes few cues from the land. The long, open streets accelerate and funnel the wind. The existing drainage pattern of the land is interrupted.

Planners force houses onto tiny plots, the setting-out preordained by the axis of the grid, not by orientation to sun, wind, and sea view at each location. Since a different set of people designs the standard house plans, it sometimes happens that the entrance porch of a house has to be lopped off so that the house may fit the plot's mandatory separation distances. The two-lane gravel street, intended to

Snowdrifts submerge village plan. Baker Lake—May

Raised beach sways the village plan. Hall Beach—April

bring trucked services to every building, dominates the plan. Planners advocate and enforce elementary zoning by-laws in the name of orderly development, and to prevent gross mixing of incompatible land uses (fuel dumps and residential areas are kept apart) at the expense of screening out other benign mixed use that might improve life in the deathly quiet parts of the community. The planners' instinct to centralize public activity means locating major buildings and commercial ventures in the geographic centre, where the encirclement of residential zones hampers future expansion. Central shopping centres—a few stores and offices under one roof grouped around a dingy blind corridor—have sprung up in many high latitude communities in Canada. People used to inhabiting large outside spaces must somehow adapt to spaces reminiscent of cave dwellings. The village has become a public housing project. Instead of anonymous apartment units in high-rises there are anonymous, detached housing units with gable roofs.

Community planning really means doing the thinking that leads to profound consensus on the purpose and direction of the settlement; it means putting words to the idea that self-governing peoples must exercise their right to live in settlements designed to complement and support their chosen lifestyle. It means thinking with the heart as well as the brain. It means having a collective memory that actively informs the future so that the community can dodge the fate of repeating the same mistakes. It means the inhabitants, who understand what is at stake, taking control.

STARTING SMALL

People and governments do not give architects the opportunity to shape whole communities. Instead they give commissions to design individual buildings or perhaps a small neighbourhood. In the absence of a profound consensus on the objectives of the community plan, how does an architect design a small neighbourhood?

In the absence of a guiding light, start small, from the beginning. First, break the mould. To repeat the stereotypes, reinforcing the impasse between existing community planning results and present and future lifestyles, serves no useful purpose. Breaking the mould means re-examination of all the parts. It means developing new moulds, as far as the budget allows, based on probable community objectives

Snowdrift analysis for community centre. Pelly Bay—April

voiced by people and businesses in the neighbourhood. It means making the project a role model, a leader in the quest for a viable community fabric. It means showing that use of outside space, building form, colour, texture, and transparency can be manipulated to improve a way of life. It means provoking thought by example, so that subsequent projects go even further in pointing the way to an indigenous community plan devised primarily to serve people rather than goods and services. It means stimulating built form to suit a range of needs and desires so that a neighbourhood identity, a sense of place, can emerge.

Second, program the outdoor space in the neighbourhood in the same way as indoor space is programmed. The spatial needs of outside activities—hunt preparations, skin stretching and drying, clothes drying, snowmobile repair, child's play, dog feeding, fish filleting and smoking, berry picking, caribou carving, and other mixed, communal and individual activities—must be assessed. Similarly, the activities that transcend the boundary between indoor and outdoor spaces—entering, exiting, waiting, casual viewing, sunning, child supervision, calling out, listening, storage of tools and materials—must figure in the assessment.

Third, abandon the notion that the neighbourhood outdoors is no more than leftover space between building forms designed by different people at different times, random space beyond the reach of thoughtful planning.

Fourth, identify the least useful, least attractive part of the site and improve it by constructing a building over it; leave the better parts exposed for the enjoyment of inhabitants and neighbours. Keep important sight lines open.

Fifth, control the scale, the apparent mass of the building, so that it does not overwhelm its neighbours and the local topography. If in doubt, choose a scale appropriate to humans walking.

Sixth, minimize disturbances of the landscape. Conserve a representative proportion of the existing flora, species and individuals that have taken generations to establish themselves. Defend the flora against incidental damage during construction.

Seventh, avoid sources of artificial noise and odour. The tundra and the boreal forest produce few noises and smells obnoxious to human senses. Noises and smells drift downwind. The best sites for inhabited buildings will be found upwind of main generators, airport ramps, fuel outlets, sewage transfer points, and garbage dumps.

Eighth, raise community consciousness of site constraints wherever the opportunity arises. For instance, place climate data including a wind rose (Figure 1-10, page 51) on the cover sheet of a set of construction drawings, and at the front of a new building's operation and maintenance manuals.

ORIENTATION

Although the decision on how to place the building on its site and orient its main axis belongs to the architect, the zoning rules often skew the issues. The building lot may only be wide enough to orient the project perpendicularly to the street, since fire control separations between buildings must be respected. Existing roads, service lines, and delivery habits worn smooth by repetition tend to define the where and how of servicing buildings.

People outdoors at high latitudes welcome free heat from the sun, the pleasure of warm sunlight received through clear, cool air. Establish a focus. A building site can be used as a sun trap by locating the building at the poleward end, so that the building's facade can absorb direct and indirect sunlight from the equatorward end of the site. A building whose long axis parallels the equator amplifies the effect. Long, cold shadows are thrown off the site toward the pole. Orienting the building slightly toward solar afternoon produces the maximum effect. Different building types mean different preferences for exposure to sunlight in the course of the day.

Building surfaces in the lee of the wind that face the sun warm up quickly, and reradiate heat inward and outward. The purpose is not to banish the effects of the wind—people at high latitudes are used to wind, and would be lost without it—but to minimize negative effects in the vicinity of buildings. It makes no sense to orient building entrances so that wind laden with rain or snow blows into the house when the door is opened.

Principal views to be preserved: the water's edge and the main approach routes of returning hunters. Other views include remote headlands, sails of stranded icebergs, whale and walrus feeding grounds, the pole star, the axis of the aurora, and the main approaches to the building or to the neighbourhood. Principal short views may include the entrance to the community centre, the food co-operative, or the school playground.

SNOWDRIFTING

Building orientation must also acknowledge snowdrifting, particularly in communities situated beyond the tree line.

Consider a large obstacle—say a school with a gymnasium and a heated crawl space—placed in a stream of blowing snow beyond the tree line with no regard for the problem of snowdrifting. Figure 1-18 approximates the air flow and snow accumulation

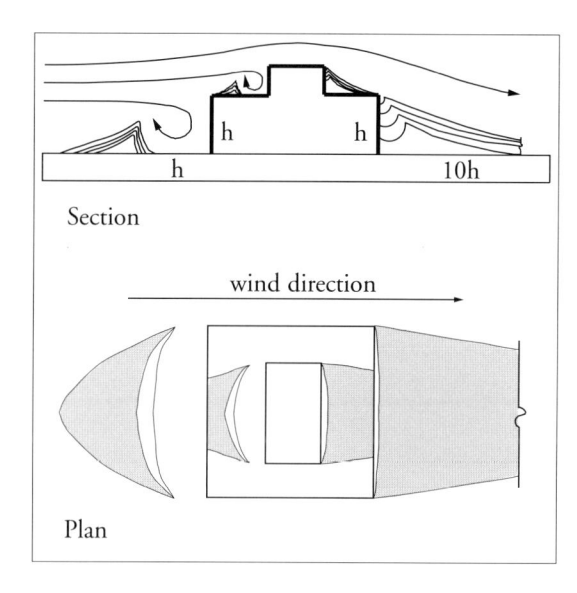

Figure 1-18 Building with closed crawl space; estimated drift pattern due to blowing snow.

around the building during a moderate winter. Of the two distinct types of accumulation the windward drift is smallest and forms a distance upwind of the building about equivalent to the height of the first major obstacle. The air speed on average in this zone is highest near the building, where the air stream splits. One part of the stream turns down, scouring the ground, losing enough speed along the way to deposit some of its cargo. This densely packed drift, seen in section, resembles a breaking wave with its sharp top edge releasing a "whitecap" or plume of blowing snow.

The base of the leeward drift is very long, about equal to ten times the height of the last obstacle. Its top surface hugs the leeward face of the building and eventually submerges the upper edge. The leeward

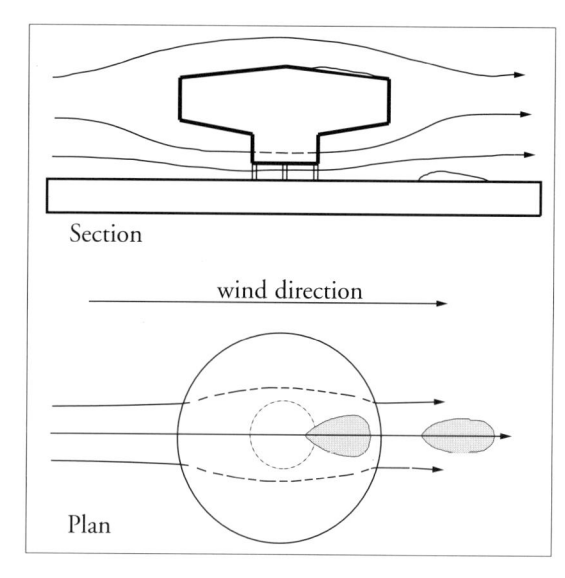

Figure 2-18 Building shaped to minimize air turbulence; estimated drift pattern due to blowing snow.

drift is packed, but less densely than its windward sibling. Where the wind accelerates, at the base of the windward wall, and around the sides of the building parallel to the prevailing wind, "wind scour" develops, often enough to keep the ground exposed all winter. This in turn has other consequences, such as raising the permafrost table and providing snow-free (but windy) entrances and exits to the building. The drifting pattern at roof level mimics the conditions at ground level.

At the other extreme, Figure 2-18 shows a small building with a shape purposely refined to cope with snowdrifting. Here the bulk of the building lies above ground level so that air flow can accelerate below it as well as above and around it. The shape does not slow the wind enough for snow particles to

96 DESIGN

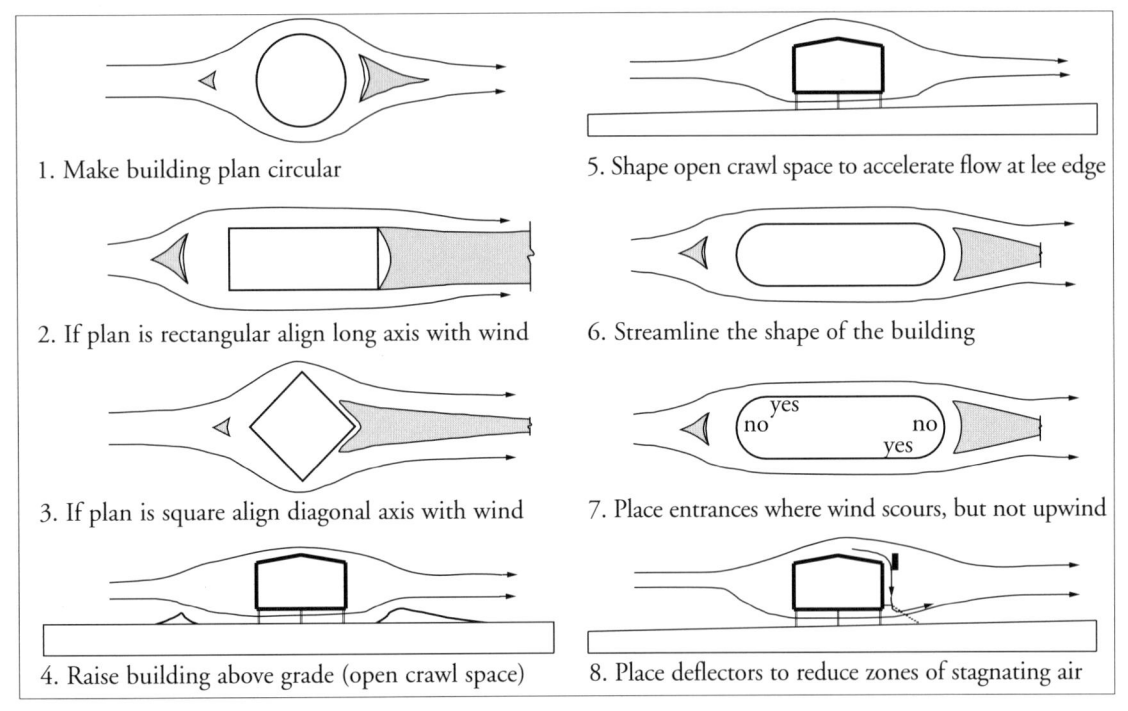

Figure 3-18 Eight rules of thumb for controlling snowdrifts around buildings.

1. Make building plan circular
2. If plan is rectangular align long axis with wind
3. If plan is square align diagonal axis with wind
4. Raise building above grade (open crawl space)
5. Shape open crawl space to accelerate flow at lee edge
6. Streamline the shape of the building
7. Place entrances where wind scours, but not upwind
8. Place deflectors to reduce zones of stagnating air

drop out of the stream. Since making buildings in the form of mushrooms often defeats purpose or practicality, forms of new buildings tend to fall between the two extremes illustrated.

Snowdrifting is site-sensitive. Where drifts will form and to what depth depends on the precise mix of wind direction, wind speed, topographic roughness, spatial relations of neighbouring buildings both upwind and downwind, the shape of the building, and the quantity of snow particles in the airstream existing at a site. Knowing precisely where snowdrifts will form is vital to building design. A snowdrift that forms next to an exit or entrance, or one that blocks the approach of a fire-fighting vehicle, has obvious consequences. Yet a snowdrift formed just two metres farther away may be tolerable. Only a scale model (of the surrounding terrain, the settlement, and the proposed building) immersed in a moving stream of water laden with fine sand particles can predict snow accumulation across a specific site with reliability.

The project may not justify the expense of a water flume trial. For projects simple in shape and scope approximate predictions of snow accumulation may be made by applying one rule of thumb. Modify the configuration of building and site to minimize the formation of large air turbulence zones that cause steep drops in local wind speed. (At the scale of a single building the aim is not to eliminate snowdrifts, but to control their size while displacing them a tolerable distance away from the building and its approaches.) A subset of rules, illustrated schematically in Figure 3-18, follows:

1. Make the building circular in plan.
2. If the building is rectangular in plan, align the long axis parallel to the prevailing winter wind.
3. If the building is square in plan, align the diagonal axis parallel to the prevailing winter wind.
4. Raise the building above the ground to permit accelerated air flow beneath the structure; in this case, if the building is rectangular in plan align the short axis parallel to the prevailing winter wind.
5. Shape the ground below the structure so that airflow accelerates most at the lee edge of the building.
6. Streamline the shape of the building; for example, keep the roof profile low, or align the roof ridge parallel to the prevailing winter wind.
7. Place the plane of entranceways and exits parallel to the prevailing winter wind, so that the adjacent ground will be scoured.
8. As a last resort, apply panels to the building that deflect the airflow into zones of stagnating air.

Several of these rules have been applied to the building seen in Figure 4-18. The long, two-storey portion of the building, which has a heated crawl space at ground level, is aligned with the prevailing wind; the stairwells at each end have been shaped to reduce turbulence in the air flow passing the end walls; the roof has a low profile. A small leeward drift results. The one-storey wing of the building projects into the air stream, but is raised above grade so that the flow accelerates below it. The leeward drift in this case begins several metres downwind of this wing, instead of against and below the lee wall. To be certain that no snow settles into the main entranceway a deflector directs wind coming over the roof straight down onto the landing. The perforated floor of the landing allows the air stream, with its load of snow, to pass through to the ground. There the accelerated air flow from below the wing sweeps the snow downwind through the open risers of the entrance stairs.

A ninth rule: advise users and maintainers how the snowdrift control features of the new building are expected to behave. An open crawl space commandeered for dead storage, or choked by stockpiles of snow cleared from the driveway, quickly surrenders its snowdrift control characteristics. It becomes part of the problem.

PRIMACY OF THE VEHICLE

To the observer on foot the existing community plan in remote settlements seems to favour vehicles over pedestrians. Vehicles can and do go everywhere at high speed, making big allowances for rock outcrops and buildings, and small ones for pedestrians. The roofs of some buildings, ramped up to the eaves with snowdrifts, are considered part of the snowmobile circuit. Many buildings sport scars and welts left by glancing vehicles. To the nervous observer vehicles appear with lethal momentum from any direction without warning other than engine noise baffled or echoed unpredictably by intervening buildings. Planning a building around the notion of a "front" and a "back" seems foolish when vehicles approach from all directions. Vehicular traffic constantly reminds inhabitants and visitors alike that things are more important than people.

If individual buildings cannot change the habits of community drivers overnight they can certainly lead by example. Vehicular traffic seeks the fastest route; pedestrian traffic the most direct route. By using the position of rock outcrops relative to the building, the form of the building itself, abrupt changes in ground level, and artificial obstacles, the planner can separate vehicular traffic from pedestrian pathways. By positive reinforcement—the placement of fixtures that favour vehicles (electrical outlets for block heaters)—random parking habits can be altered to isolate vehicles in a designated part of the site. On the coldest days parked vehicles with engines left running to avoid a hard start produce a noxious ice fog best left downwind.

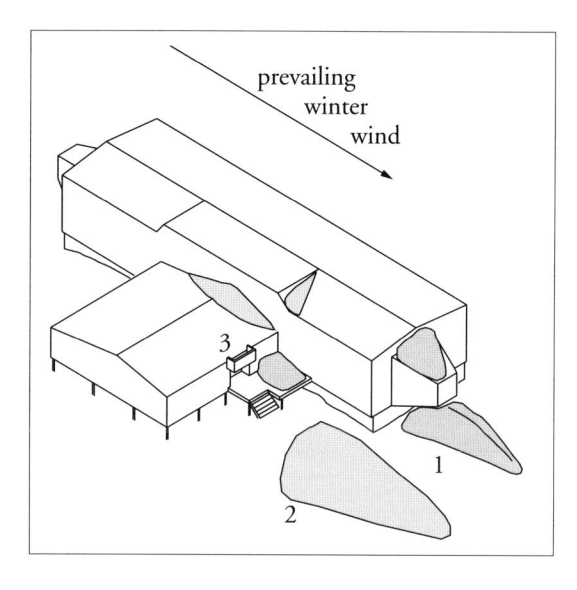

Figure 4-18 Main results of a building form designed to control snowdrifting: small drift (1) in lee of wing aligned with prevailing winter wind; shallow drift (2) displaced downwind of wing raised above ground; deflector (3) clears snow from leeside entrance. Without controls snow would drift up to the leeside eaves.

GEOMETRY

Highrise apartments facing the sea. Nuuk—August

TRADITION

People who have to make their own tools develop an acute sense of the possibilities of form. Aboriginals invented and produced clothing, shelter, tools, toys, games, and amulets—everything necessary to make life on the move both practical and rewarding. Anything that could not be carried on the back or in the mind was useless. Anything that could not be used within the confines of a small space had no future. Preference was given to games requiring play in turns, or in pairs—cat's cradle, thread the needle, neck pull, leg wrestling—not to those needing a level field and team play.

Traditional home-made tools are not hit-and-miss inventions. They are samples of a pattern improved and perfected by countless generations of toolmakers. A new example of an old tool, even if its workmanship is only fair, will be successful. Still, it is not a blind copy. Familiar with the tool's use, the toolmaker knows precisely which points are critical to its function. The special care lavished on those points makes the tool a tool and not a toy. Tools extend and concentrate human effort toward special ends. The economic use of form depends on understanding the geometry of forces, how to direct small forces to large effects.

PROBLEM

Industrial society delegates tool-making to specialists who design machines to design and manufacture tools. Tool-making is no longer there to provide the majority with a basis for intuitive understanding of the use of geometry, for the skilful use of less to produce more. Building designers are no exception: the economic use of form is not part of their training either. Their clients are also in the dark.

In practice, giving form to building projects is a semisecretive process rarely exposed to rational analysis. The process is kept semisecret precisely because it is not very rational, even though the end product itself may be given a rational basis for being. Who wants to volunteer the information that the design thinking crucial to the project is not rational? The expectations surrounding the concept "rational" are overwhelming. People expect form-giving to be a deliberate, deterministic, step-by-step search for a positive result. Designers agree that the deliberate approach is essential to the process—subject to intuitive, sometimes decisive interventions of the irrational part of the brain. By sidestepping long searches down alleys that usually turn out to be blind, the irrational part of the brain is free to jump to conclusions. It can then perform simple tests on such conclusions, dismiss those obviously of little use, and return the possibly useful remainder to the analysis of the deliberate approach.

Economy of form is not the first rule of the design process. Other important determinants of form exist to breathe life into the project. Still, it happens too often that the designer fails to detect those determinants or, aware of them, fails to respond to them adequately. A design solution that responds appropriately and with imagination to the project determinants does not emerge. It is a bad design even if it is demonstrably safe to occupy. But is there any excuse for a bad design to be uneconomic in form as well? The owner faces accepting a building that not only falls short of his hopes but also costs more to build and operate.

PROPORTION

The size of shelter built by humans has traditionally been inversely proportional to the harshness of the out-

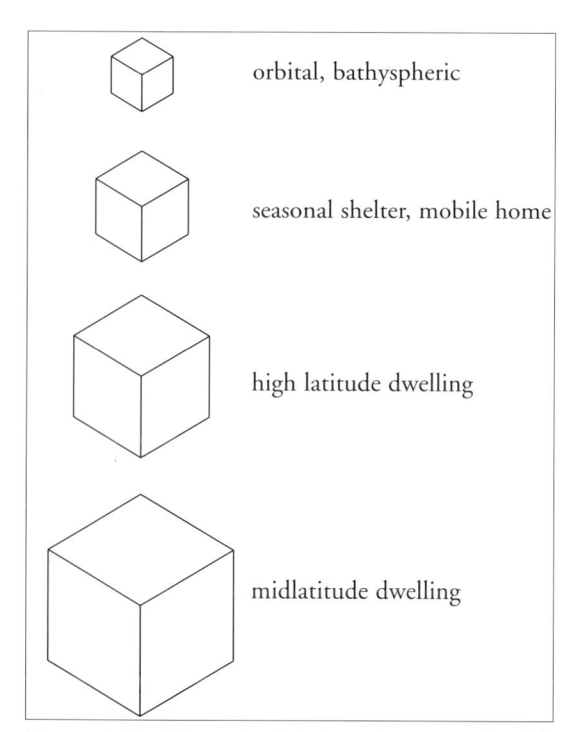

GEOMETRY 99

Wind deflector at school building. Sanikiluaq—April

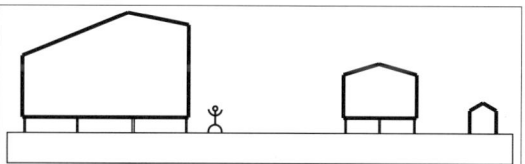

Figure 1-19 The scale of shelter: the greater the difference between outside and inside conditions, the smaller the shelter.

side environment. The greater the difference between outside and inside conditions, the smaller the shelter. Figure 1-19 indicates that a bathysphere designed for a deep undersea environment is much smaller than a shelter made for a suburban environment. A high latitude shelter falls between these extremes. The snow house is an obvious example. Available resources have to be concentrated in a small, robust form. Concentration of resources in a small form means that everything contained must have its place, must fit in, must permit the activity of the moment to occur without hindrance. Space inside a small shelter is precious.

Figure 2-19 Remote settlements once dominated by tent-sized objects that hugged the tundra now consist of gymnasium-sized objects suspended in the air. Humans used to stand tall by their shelters; now they seem like dwarfs.

Large buildings in northern Canada, particularly those built over ice-rich soil, are often raised a metre or more above ground, leaving an open crawl space. Add to this at least two metres for a heated crawl space that includes floor structure, and the building towers over the humans that use them. (In the tropics many traditional shelters are set on tall stilts, permitting use of the space below, so the problem of human scale does not arise in the same way.) Add imported concepts like a "gymnasium" whose ceiling must be a minimum of six metres above the floor and the problem becomes more extreme: the roof may be ten metres above the ground (Figure 2-19), more than eight metres above the heads of passers-by.

PERIMETER-TO-AREA RATIO

Floor area is the basic measure of a modern building; owners want as much usable floor area as their budget will allow. The shape of the floor plan—the footprint—determines how much floor, wall, and roof must be built to accommodate the required floor area. Economical building forms require the least wall perimeter to enclose the most floor area: a low perimeter-to-area ratio. Figure 3-19 shows five sample floor plans, each having the same floor area: four units. The circular plan, with a perimeter-to-area

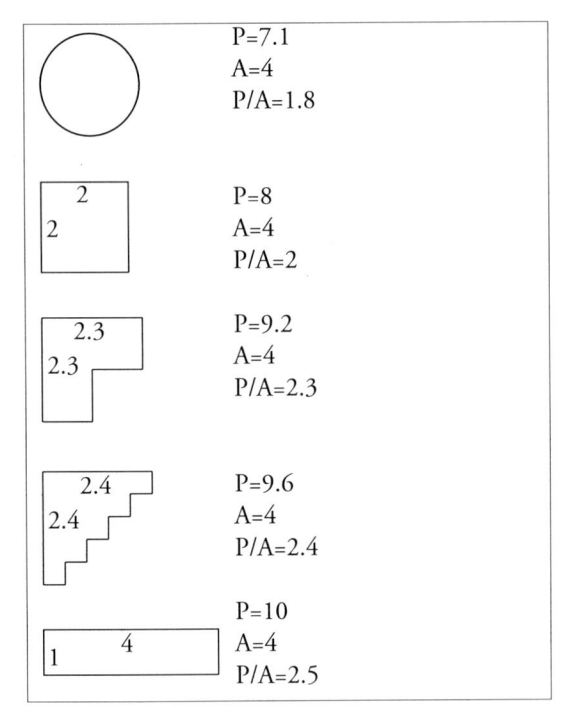

100   DESIGN

```
○        P=7.1
         A=4
         P/A=1.8

 2
2        P=8
         A=4
         P/A=2

2.3
2.3      P=9.2
         A=4
         P/A=2.3

2.4
2.4      P=9.6
         A=4
         P/A=2.4

 1  4    P=10
         A=4
         P/A=2.5
```

Figure 3-19 Five building plan forms having the same floor area (4 units) but very different perimeter-to-area (P/A) ratios. The longer the perimeter wall the higher the construction/operating costs.

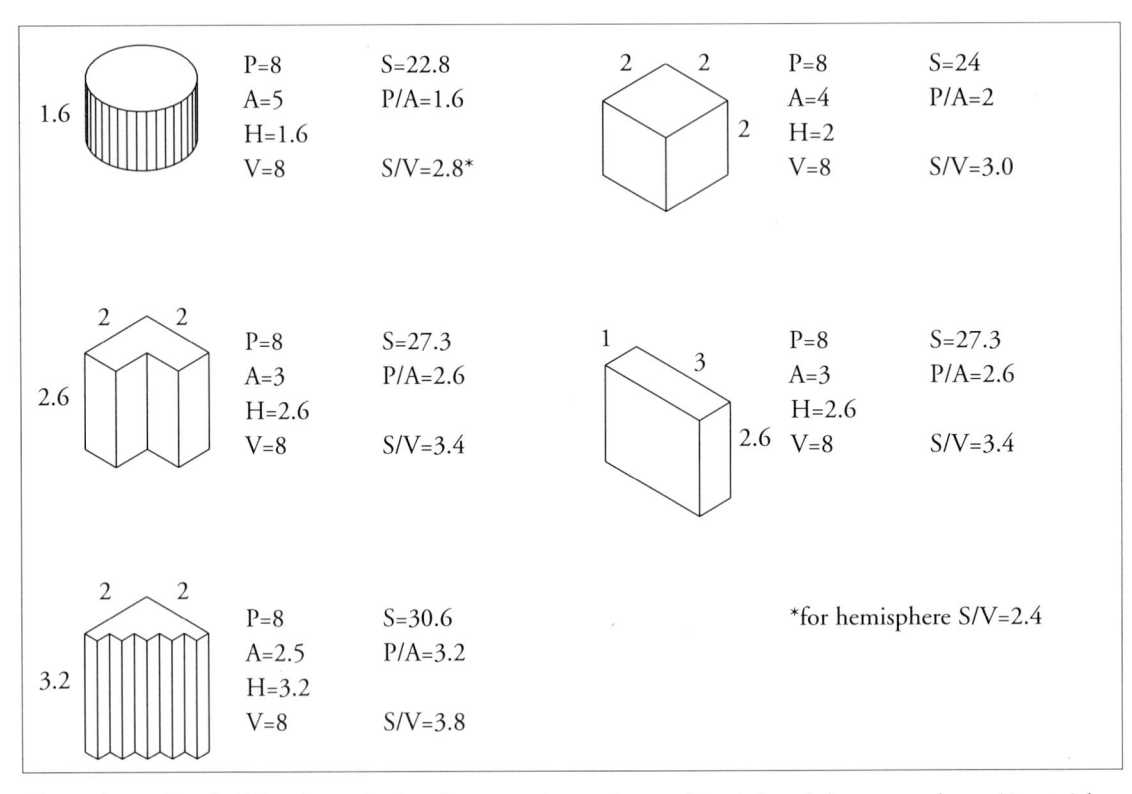

```
1.6   [cylinder]    P=8      S=22.8       2  2  [cube]    P=8     S=24
                    A=5      P/A=1.6                      A=4     P/A=2
                    H=1.6                          2      H=2
                    V=8      S/V=2.8*                     V=8     S/V=3.0

2.6   [L-prism]     P=8      S=27.3       1  3  [slab]    P=8     S=27.3
                    A=3      P/A=2.6                      A=3     P/A=2.6
                    H=2.6                                 H=2.6
                    V=8      S/V=3.4             2.6      V=8     S/V=3.4

3.2   [finned]      P=8      S=30.6
                    A=2.5    P/A=3.2
                    H=3.2
                    V=8      S/V=3.8
```

*for hemisphere S/V=2.4

Figure 4-19 Five building forms having the same plan perimeter (8 units) and the same volume (8 units) but very different surface-to-volume (S/V) ratios. The forms with the smallest ratios are the least expensive to heat.

ratio of 1.8, encloses the given area with the least length of wall. The plan having one very irregular side requires a wall forty percent longer to enclose the same area.

SURFACE-TO-VOLUME RATIO

Heat loss, or heat gain, of a building is proportional to the total surface exposed to the outside environment. For like-sized building forms the greater the surface exposure, the greater the heat loss.

Similarly, the endurance of a short, compact human body in cold weather usually exceeds that of a tall, thin one. In order to lose more heat, a radiator's total surface area is increased many times by the addition of fins. Economic building form at high latitudes retains heat energy as long as possible by having a low surface-to-volume ratio. A building that resembles a radiator, like the one in Figure 4-19 that has a surface-to-volume ratio of 3.8, would be comparatively expensive to heat at high latitudes. (A building with balconies attached to its structure without *thermal breaks* behaves even more like a radiator: the surfaces of the balconies represent a huge increase in the total surface area of the building exposed to the outside.) A hemisphere of similar volume, with a surface-to-volume ratio of 2.4, can be heated much more cheaply.

In building forms of dissimilar size, say a detached house and a small community centre, the surface-to-volume ratio of the former is greater than

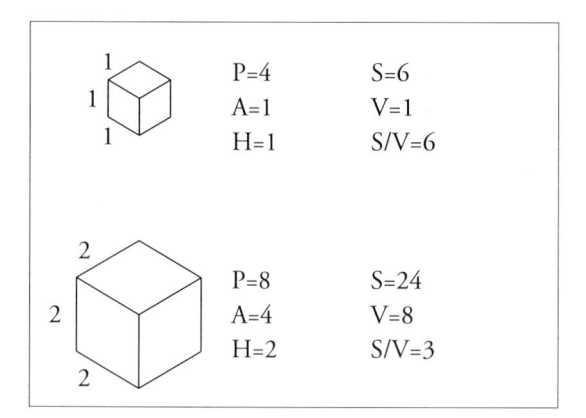

GEOMETRY 101

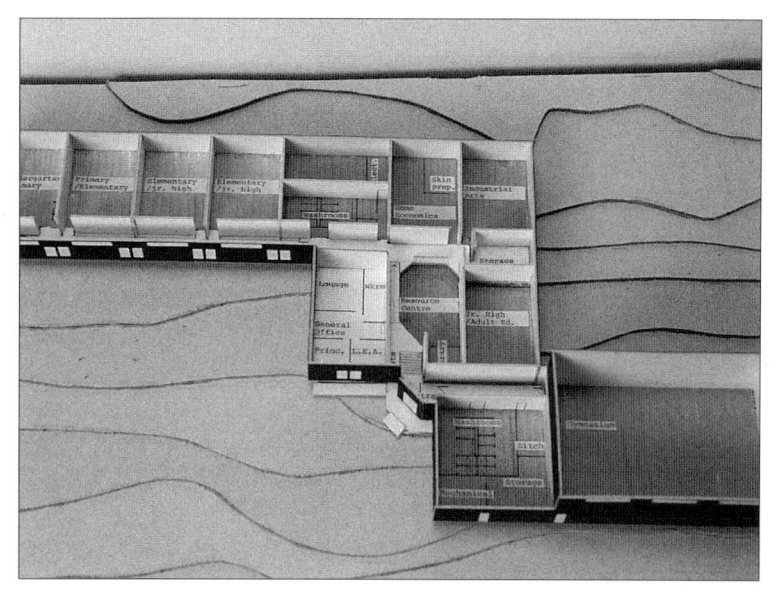

Programming model for new school. Lutselk'e

Figure 5-19 Since the surface-to-volume ratio is greater for small volumes compared to large, a small detached house represents maximum surface exposure to outside conditions.

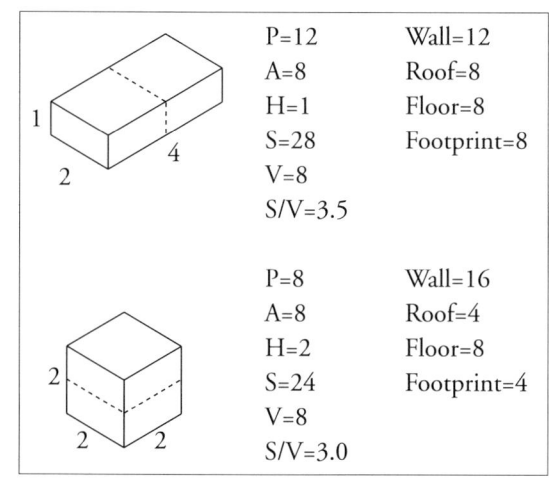

```
1         P=4      S=6
  1       A=1      V=1
 1        H=1      S/V=6

2         P=8      S=24
  2       A=4      V=8
 2        H=2      S/V=3
```

```
           P=12     Wall=12
1          A=8      Roof=8
           H=1      Floor=8
  2   4    S=28     Footprint=8
           V=8
           S/V=3.5

           P=8      Wall=16
           A=8      Roof=4
  2        H=2      Floor=8
           S=24     Footprint=4
  2  2     V=8
           S/V=3.0
```

Figure 6-19 Two forms of detached housing: the single-storey version has the same floor area and volume as the two-storey one, but exposes 15 percent more surface area to outside conditions.

that of the latter (Figure 5-19). The smaller the volume the greater the influence of the surface, and therefore of heat loss, on that volume. Detached houses represent maximum exposure to the outside.

At high latitudes, consideration of surface-to-volume ratio must figure in any debate about the virtues of single-storey house construction versus two-storey. Figure 6-19 shows that a single-storey house has sixteen percent more exposed surface area than a two-storey house of same volume. Expressed as heating dollars, this difference looms large over the lifetime of a house. The two-storey house also offers a half-size footprint—a significant discount where the cost of developing land and installing foundations is high.

CONCENTRIC ENVELOPES

A modern building is essentially a heavier-than-air, semi-rigid balloon. Like a balloon, the building buckles or sags under the impact of forces such as wind, earthquake, and gravity. Building codes regulate the degree to which a building may buckle or sag without endangering the people within. The building's skin or envelope must be continuous, despite the piecemeal manner in which humans are obliged to put larger-than-self things together. Design and construction of the building envelope cannot succeed without grasping the concept of continuity: a punctured balloon is no longer a balloon.

Envelope continuity is usually easier to achieve outside the building form than inside. Continuity inside is interrupted where internal structural supports meet exterior structural framework.

JOINTS

Buildings are made of countless parts manufactured at different points of the globe, and the finished whole is expected to be adaptable to future needs. Casting a building in one piece, even if it were feasible, would

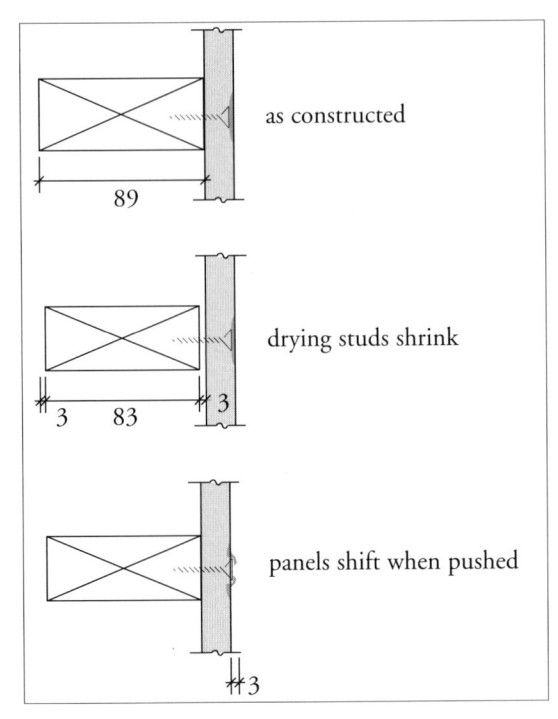

Figure 7-19 Post-construction shrinkage of wet or "green" wood studs reduces back support for wallboard panels. Wind or contact pressure shifts panels toward studs. Screw heads pop out.

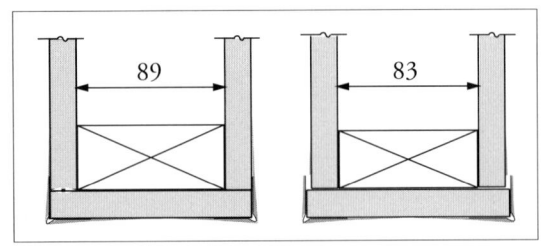

Figure 8-19 Post-construction shrinkage of wet or "green" wood studs causes cracking in finish surfaces where there is some resistance in the direction of movement.

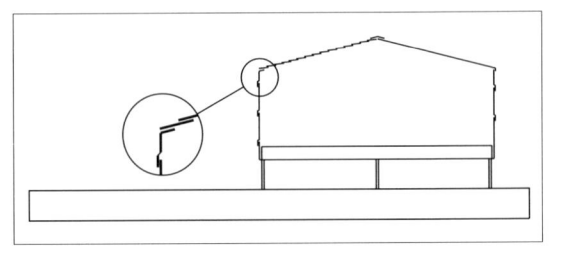

Figure 9-19 Lap joints in building envelope components are usually configured to shed water in stages to the outside, away from the building interior.

diminish the flexibility that distinguishes human shelters from bat caves. Parts come together at joints. Without joints that work properly there is no whole. Individual parts can be described in a few pages, but showing how these parts come together takes many pages and many drawings, the bulk of the project's construction documents. Unless the type, size, location, and frequency of joints is planned meticulously in advance, one part of the building will antagonize another, causing costly breaks in envelope and structure.

The frequency of joints in a building form is a matter of geometry and choice. One school states, "Joints are problematical: the fewer the joints, the fewer the problems." The other states, "The more joints there are, the less the work any individual joint has to do, the fewer the problems." The material in a given length of wall, for example, will contract a predictable amount at a low winter temperature irrespective of the number of joints: the total movement must be accommodated by either a few wide joints or many narrow joints. Reducing the number of joints just relegates the total movement problem to fewer places. This means that the design of joint and weather seal must be of high quality. It also means that large parts or panels must be of a quality high enough to resist shipping, installation, and service conditions with minimum support.

Compare vertical wood siding in 100 and 150 millimetre widths, both having high moisture content when installed. The tongues and grooves of both types are dimensioned alike. As the two types dry out they shrink; tongues retract from grooves a small distance proportional to the total board width. Although there are fewer joints using 150 millimetre siding, each board shrinks so much that tongues no longer engage grooves.

The first loyalty of building parts is to themselves, not to the assemblage or whole. Each part behaves differently than the next in response to changes in humidity, temperature, ultraviolet light, differential settlement, change of plane, and the like. Building parts are inherently weakest at the edges.

In northern Canada the commonest examples of unplanned joints occur in drywall construction. Relatively dry, heated interiors shrink the wood studs in walls and partitions more than the plaster wallboard attached to it. Add to this movement the strain of differential settlement in the foundation, however slight, and the plaster wallboard will crack in lightning bolt patterns.

Two related problems in wood frame and wallboard construction are illustrated in Figures 7-19 and 8-19. In the first, the wood stud shrinks so severely across the grain that the distance between the head of the drywall screw and the face of the stud increases measurably. The panel remains suspended on the shank of the screw but has lost its backing. Even small amounts of pressure applied by wind forces or

GEOMETRY 103

passing humans push the panel towards the face of the stud so that the screw head bursts from its plaster cover. In the second example, a wood stud occurring at the free end of a partition shrinks. In shrinking it sucks the wallboard of the opposing faces together while the wallboard at the end face, lying in the direction opposed to the direction of shrinkage, resists the strain.

LAPS

To ensure continuous watershedding action at joints, roofers install roofing material such that upper layers always lap lower layers. In the case of asphalt shingles or sheet metal panels this is the only way to achieve waterproofness. Any other material that may be exposed to rain, meltwater, or water from ruptured pipes should also be lapped at horizontal joints in such a way that any liquid running inside a wall or roof is deflected toward the outside of the building, not the inside. Figure 9-19 illustrates this ancient principle.

SEALANT SHAPE

Sealants manufactured to fill joints in buildings are not designed to accept joint movement in more than one axis. Despite this, installers routinely squeeze sealant into the bottom of joints so that the material adheres to more than two opposing surfaces, creating the basic condition for premature joint failure. The sealant becomes elastic shortly after installation; its shape, its cross section inside the joint, is crucial to its durability. Figure 10-19 shows shapes that endure and some that fail early in the life of a building. The ideal shape adheres only at opposite sides of the joint. Both the inner and outer face of the sealant bead are convex. The inner convex face is formed against a

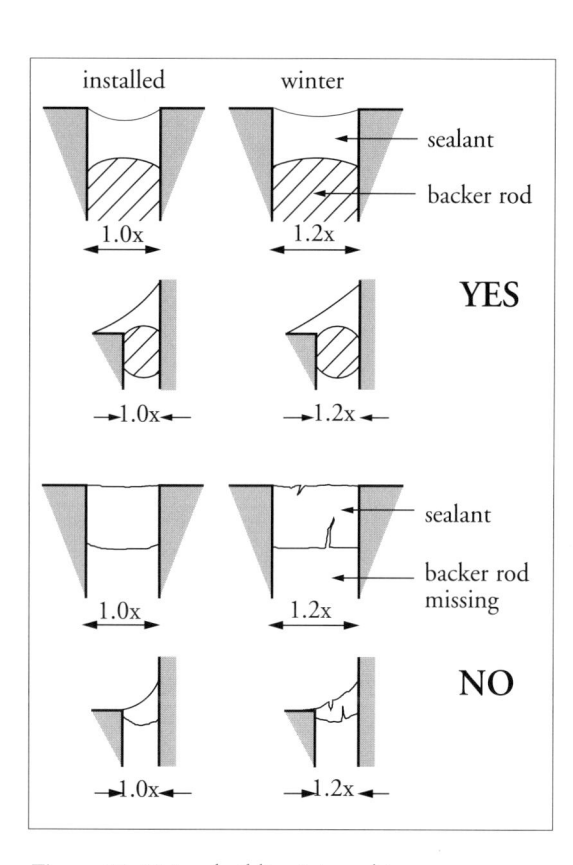

Figure 10-19 In a building joint subject to movement the shape of the sealant in cross section is as important for durability as the quality of the material.

friction-fit backer rod made of flexible foam plastic. If the joint is shallow a strip of bond-breaking material must be placed on the bottom so that the inner face of the sealant bead is at least free to move as much as the outer face.

At high latitudes ideal conditions for installing sealants are rare. In winter not only is it difficult to install material so cold that it is more solid than fluid, but the joint itself is at its widest, given the contraction of the adjacent panels. Sealant installed in winter will be over-compressed in summer when panels expand, narrowing the joint. Adhesion to wet/frozen building components will be poor.

INVERSE SQUARE LAW

The intensity of a point source of light or sound decreases for the viewer or listener as the square of the distance between the point and the receiver increases (Figure 11-19). The intensity of a light bulb in a darkened room decreases geometrically for the viewer who walks away from it. Similarly, a speaker's voice drops off rapidly in the ear of a listener who walks away.

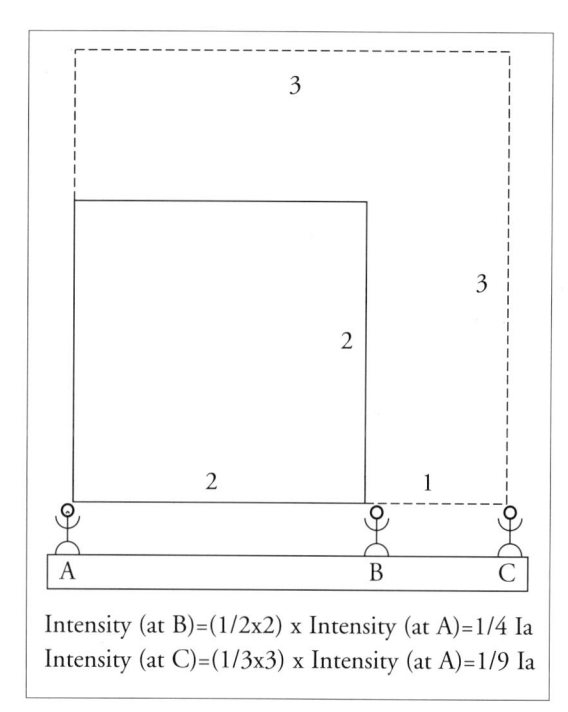

Intensity (at B) = $(1/2x2)$ x Intensity (at A) = $1/4$ Ia
Intensity (at C) = $(1/3x3)$ x Intensity (at A) = $1/9$ Ia

Figure 11-19 Inverse square law.

The inverse square law becomes especially important when planners decide to compress space in order to save money. If people in a partitionless office are currently spaced at head-to-head distances averaging 2.8 metres, a proposed reduction of the average distance to two metres may seem unimportant. But any sound emanating from one person has now doubled in intensity in the ears of the next person. Left unchecked this increase in noise level will diminish the comfort of people at work.

SCALE MODELS

Scale models of new building projects usually come into public view when the design process has come to an end. They represent the product of a thousand decisions made out of public view; they represent a take-it-or-leave-it proposition.

Scale models made during the design of the building and shown to client and user at intervals represent more than an end product. They are a tool in the development of the project. They provide a sharp focus for discussions between client, user, and designer that, despite the advent of computer graphics, has no equal. Advantages and disadvantages of the current model tend to leap out at the viewer, regardless of the viewer's background or lack of background in the development of building projects. Subsequent models show the input of previous project meetings clearly. Client and user feel more like involved human beings and less like token participants.

TRADITION

People who are inseparable from the land, who depend on it for sustenance, do not lose their way.

PROBLEM

Startled awake from a deep sleep humans may experience a flash of panic punctuated by the thought, "Where am I?" Assuming that nature puts first things first this panic signals the importance of knowing exactly where a person stands in the known universe. Am I safe?

Typically we pinpoint our location by means of spot checks of familiar sights and sounds around us. Landscape we learn to know, and shelter built with our own hands affirm our sense of security. People who speak greedily of waiting out a storm in a warm, dry place, or of curling up in bed with a good book, speak of security. In contrast, a building designed and built by strangers raises an awkward screen between people and the familiar points of their landscape. A badly designed screen undermines one's sense of security; a well designed one amplifies it.

Knowing the way—finding one's way—inside any building at any latitude depends partly on visual contact with the outside. At high latitudes long periods of darkness in winter and small windows in buildings tend to frustrate such contact. In a large school with few windows, or with windows raised above eye level, finding one's way must become a conscious effort. Frame of reference is eliminated. Look-alike corridors force students to watch or listen for other cues—a number on a door or a teacher's voice—on the way to finding the right classroom. People and rats can adapt to a labyrinth, but should they have to?

Finding the way depends also on the clarity of the building's purpose as conveyed by the scale, shape, and continuity of the spaces encountered. Clarity means that the building's spaces suggest when to move, to stop, to rest; it means recognizing arrival intuitively; it means recollecting the way out intuitively. Beginning with the stress of "Where am I?," lack of clarity in the building plan may end in the frustration of "Why do I bother coming to this place!!?"

CIRCULATION

In many buildings at Canadian high latitudes circulation space is sacrificed to obtain the maximum usable or rentable space. Circulation space, seen as a necessary evil, may not even rate a mention in the building program except as a slim percentage of the projected gross area of the building. People in remote settlements have no choice but to visit a public building to ferret through dark corridors for the service they seek—they are a captive market. Being lost in a building, hastening to the end of another corridor only to find a new set of ambiguous choices, is a common experience that indicates that, for the owner, the user's sense of well-being is not a priority. The circulation space is "left over" from the exercise of designing and aligning the main rooms, just as the exterior space between buildings is "left over." Like antimatter or the far side of the moon it exists only because nature demands back ends to things.

A tree gathers nutrients from the ground and passes them through trunk, branches, twigs, and leaf stems to make leaves at the sunlit periphery. Gibbons feeding or posturing in the tree know their position relative to the trunk by the diameter and suppleness of the branch supporting them. Signposts are unnecessary. In well-designed buildings the main entrance gathers up people and, through lobbies, corridors, and aisles, directs them to their destination. Good circulation space broadcasts the purpose of the building and the scope of its activities, making signposts and ushers unnecessary.

Different buildings require different treatments of circulation space. Common to all, however, is a sense of progression, either large space to small, public space to private, general-use space to special-use, or interdependent space to independent, so that at every step one's position relative to the whole can be understood without a map. Every step from first sight of the building to arrival at its core seems logical, and intentional rather than accidental. The visitor and user are equally comfortable, never far from contact with the outside through sight lines, sounds, daylight or sunlight. The route of travel remains short and the direction of travel never doubles back. The character of each leg of the route conveys a message: "follow through," "straight ahead," "wait here," "destination in sight," "speak to the receptionist."

The purpose, size, intimacy, duration, and timing of pedestrian flow through the building must be understood thoroughly at the program stage to prevent "accidental" circulation space becoming a reality.

Skylight over school lobby. Sanikiluaq—April

ignored by dint of imagination; hard boundaries bruise shins and limit possibilities.

The building program, chapter and verse, may seem to suggest a hard boundary for every space categorized, a set of pigeonholes. Pigeonholes may mean something in the hotel industry but not in normal social activity. The architect probes the building program for potential soft boundary sites, for places where activities can co-exist or prosper in adjacent spaces. A room housing a supervised activity requires the presence of a supervisor. Two rooms with moderately supervised activities could co-exist with a half-height partition topped with clear glass so that a single supervisor can work one room while monitoring the adjacent one and move between rooms through a common door as the need arises.

SCALE

In polar regions the high cost of construction keeps a short leash on the size of projects—most buildings are small compared to their relatives at lower latitudes. This could be a good thing. Since the design of most buildings comes primarily from the mind of one person, that mind spreads itself thinly on large buildings and generously on small ones.

Small scale can foster ingenuity, the will to do more with less. If a space can be used for two types of activity rather than one, the cost of constructing a second space can be avoided. Although a basketball game in a gymnasium can exist without an audience, space for seating an audience would stimulate play and community interest; but the cost of adding bleachers and another third to the volume of the gymnasium might deafen the ears of the owner. Instead, the owner could agree to place the main lobby and the gym together, so that a sliding wall panel between the two would allow spectators to see an important game from the lobby. The lobby becomes a viewing gallery for the gym. In this case both the gym and the lobby are less than perfect taken separately, but together the whole works better than the parts.

BOUNDARIES

Architects define boundaries to create useful space. A boundary can be almost anything: a line painted on the floor; an imaginary line on the floor (between two columns); one-way glass; clear glass; drapery; folding panels; doorways; full-height windows; half-height partitions; walls; a change of material; a change of colour; a change of pattern; a change of level; a change of height; a change of direction; a change of acoustics; a change of view; a sign; a laser beam connected to an alarm; a retractable grille; a furniture layout. Soft boundaries can be respected or

STRUCTURE

In settlements dependent on annual resupply by sea, building projects with large or complex structural systems cannot be enclosed in the brief period between the arrival of materials and the onset of winter. Polar summer is too short. Snow drifts through the frame work, shutting down work until the following spring. Buildings with simple, integrated structural systems proliferate at high latitudes for this reason. The simplest so far, the loadbearing wood stud wall, uses the same studs to support gravity loads and stiffen wall panels. Loads are evenly distributed throughout the system. Fabricating a wall panel on the floor and tilting it upright combines structure and basic enclosure in one operation. The density of the material in the exterior envelope approximates that of the structure, so that envelope and structure behave similarly under conditions of thermal stress.

Other common structural systems—post and beam, portal frame—concentrate loads at intervals. The structural members, few in number, have to be much denser than the envelope to support the concentrated loads. Being denser, these structural materials react more than many envelope materials to changes in the thermal environment. They expand and contract more when temperatures rise or fall; they conduct heat more quickly. When the structure moves more than the envelope attached to it, the potential for tearing of the envelope increases.

Post and beam or portal frame structural members are often thicker than the envelope. Where then will the structural member be placed in relation to the envelope? The choices are these: completely outside; completely inside; interrupting the envelope completely; interrupting the envelope partially. Completely outside, the structure reacts significantly to exterior temperatures, which vary as much as 80°C. Severe differential movement must be accommodated in expensive, possibly non-standard fixtures needed to support the more thermally-stable envelope. Completely inside, the structure and the inner face of the envelope remain at about the same temperature most of the year. No heat loss attributable to the structure occurs. But the thickness of the inside structure may interfere with the usefulness of the enclosed space.

Interrupting the envelope completely makes the structure react to two environments, the severe outside one and the stable inside one. The resulting movement may be unpredictable in detail. Any such break in the continuity of the barriers in the envelope increases geometrically the potential for air, vapour, and heat loss. Poor workmanship at such intersections combined with differential movement and high heat loss through the structural member (a thermal bridge) make this configuration unworthy of consideration. Yet some people still build this way. Interrupting the envelope partially rather than fully does not improve the situation by much. The bulk of the structural member exists in the stable inside environment, but its outer edge extends into the thermally unstable outer portion of the envelope. Significant thermal movement and heat loss occur, and the structure breaks the continuity of several barriers in the envelope. Regardless of the structural configuration selected, significant heat loss through structural members to foundations dependent on stable permafrost regimes must be prevented.

Load calculations for structures built at high latitudes present no surprises. Incidence and duration of live loads may sometimes be underestimated, however. Gusts of hurricane force occur at irregular intervals in many communities located near treeless, mountainous terrain. Gale force winds may persist in some places for days, exposing the structure to vibration felt and resented by inhabitants. Movement due to cyclical loading and unloading of storage tanks supported by the structure must also be considered.

Portal frame. Ilulissat—August

FLOOR

People who adapt to almost any hardship never get used to the shock of walking on cold floors in bare feet. Warm air generated by the central heating system tends to rise and stratify against the ceiling, leaving cooler, heavier air to pool on the floor. Cold air from the outside rises through air leaks in the ground floor to replace the volume of warm air lost to the outside through leaks in the upper parts of the building. Ceiling temperature might be +25°C while the surface of the ground floor remains at +10°C. The physics of air circulation seem to preclude warm feet.

Student residence dining room. Rankin Inlet

Doubling the amount of insulation in a cold floor has little effect: the problem is one of cold air infiltrating the floor and pooling, more than radiant heat loss downward through the envelope. Enclosing and heating the crawl space below the floor does raise floor temperature to comfort level, at considerable expense. This strategy fails, however, wherever heat loss from the floor must be prevented from degrading the permafrost in the soil supporting the foundations, and wherever snowdrift control demands a crawl space open to the elements. It may be preferable to install a radiant heating system just below the floor, set to near room temperature.

A double floor, an insulated outer floor supporting an uninsulated inner floor open to room air circulation through floor vents, does raise the temperature to acceptable levels. Successful floor design requires the installation of a fully operational air barrier system, an assembly impervious to air flow and resistant to air pressure. Easier said than done. Designers frequently ignore the difficulty of applying air barriers from below. Much of the year the wind makes the open crawl space extremely uncomfortable to work in. Air barrier design in raised floors must recognize the difficulty by simplifying the installation sequence.

Cold air infiltration through ground floors having open crawl spaces restricts other choices. Waste water and water pipes installed between the joists of raised floors freeze in winter on contact with cold air drafts. Floor level changes useful to the designer in defining boundaries between spaces increase the risk of barrier discontinuities just by being irregular and therefore less familiar to the builder. Floor joists with top edges in a heated environment and bottom edges in a cold one may react by deflecting upward; in time this cyclical movement will damage floor finishes.

In floors, air and vapour barriers pose a special problem: they may not be self-draining. Any water spilled on the floor settles into the valleys of a polyethylene vapour barrier, for instance, raising rather than lowering the presence of vapour in the floor assembly until completely evaporated. With a discontinuous vapour barrier membrane, the flood would drain straight through the floor assembly to the ground, causing minimum damage.

STORAGE

Although comparatively small, buildings at high latitudes need large storage spaces to accommodate once-a-year bulk shipments of food and supplies. Similarly, shelter must be found to store materials and possessions that might otherwise disappear under a snowdrift for eight months of the year. Storage is vital to polar lifestyles. Requirements for it should be understood in detail at the program stage.

21 ENTRIES

TRADITION

Snow houses built to last a month or more have tunnels for entrances. Several metres long and just high enough to crawl through, they perform several functions. Most important, the tunnel works as a vertical air lock. Its ceiling height remains at or below the level of the sleeping platform, and so prevents cold outside air from rising into the living space and disturbing the warm air, heated by human bodies and an oil lamp, trapped under the dome.

The tunnel's length and position separate the living space from snow infiltration and wind noise. Hunting gear that would be buried outside in a snow storm can be stored temporarily inside the tunnel. The tunnel provides a gradual transition between outside and inside: the insider can anticipate by a few seconds the arrival of the visitor. Eyes begin adjusting to changed brightness levels. The process of bending, crawling through the tunnel, and then standing up inside the living space emphasizes the outsider's sense of arrival in a secure place. Both snow tunnel and snow house are intimately connected with the land: on entering there is no leaving the surface of the ground to climb into a shelter raised an arbitrary distance above ground. Like the Bedouin grandfather who pitches his tent next to his grandson's bungalow, people who live on the land cannot feel comfortable in a living space divorced from the earth.

PROBLEM

Only sixty years ago a one-room cabin with a single door seemed to be a luxury to many northern Canadians. Finding the entrance to a one-room cabin, except perhaps in a blizzard, did not require special intuition. But the single entrance door was also a nuisance. It jammed continually, let in a steady stream of cold air, and frequently broke away from its hinges. Today, with most such cabins fading into the landscape, the entrances of new buildings are expected to perform basic functions well, while communicating the character and purpose of the building to visitor and occupant alike.

Many entrances to buildings in northern Canada make only one concession to the climate, the presence of a wind porch. Often just large enough to shelter the swing of the main door, these cubicles break the force of the wind. But they also fill up with snow until the outer door can no longer be closed and with, for lack of a better place, a miscellany of possessions that collect underfoot. Worse still, many roofs pitch snow, rain, and meltwater onto entry steps and landings, making safe footing impossible.

STRATEGIES

A well considered entrance sends the signal that the rest of the building may be just as well considered. The key strategy is to promote the status of the entrance from afterthought, a miscellaneous circulation space, to parity status with other important spaces inside the building. The entrance, and the space allowance for it, must have standing in both the program and the construction budget.

Character: Make the entrance highly visible from a distance and distinct from the mass of the building. Position is crucial: an entrance hidden around a corner first confuses and then frustrates the user. A newcomer should be able to decide in advance where to find the entrance and how best to

Snow-free entrance. Iqaluit—April

Snow-bound entrance. Repulse Bay—March

approach it. Distinctness can be in size, shape, shadow, light, translucency, texture, or colour. The entrance stands for welcome and shelter, and should send that message to every user, day and night, even during a snowstorm. If there are several entrances and exits to a building, the main entrance should clearly predominate.

Transition: The connection between outside and inside circulation should be obvious; outer space should lead to inner space without stops or abrupt changes of path. But the user should be made conscious of the transition, of the sense of arriving at or departing from a significant place. Changes in level, direction, and light level are most frequently used to send the message.

Access for the handicapped: At least one entrance to public buildings must be accessible to people with restricted mobility. A vertical air lock at such an entrance may be out of the question. Since buildings are commonly one or two metres above grade, the low-sloped ramp becomes a major factor in the design of the entrance. Ramps installed parallel to and near the lee face of a building cannot be kept free of drifting snow. A long gravel ramp up to entry level, laid parallel to the entry side of the building but set several metres away from it, works well in some situations (Figure 1-21). The gap between the top of the ramp and the entry threshold is bridged by a perforated steel deck or grille, which limits the amount of dirt and snow tracked inside.

Door bell: In northern Canada doorbells have been treated as a pointless luxury. To the person inadvertently locked outside during a blizzard a doorbell with which to arouse the occupants seems a necessity.

Light and translucency: Plentiful daylight and sunlight (when available) together with views to the inside and to the outside reinforce the sense of welcome. An entrance flooded daily by equatorial sunlight has a "shorter" winter than any shaded part of the building. It seems shorter to the user because sunlight moderates the microclimate, and snow and ice adhering to walking surfaces disappear earlier in the spring. Consider placing sidelights or windows near the doorway as an alternative to transparent doors, so that people can peer outside without blocking traffic.

Wind and drifting snow: To be comfortable in winter an entrance must be protected against the polar wind. However, placing an entrance in a windless location, the lee of the building, risks blockage by snowdrifts, particularly in settlements located

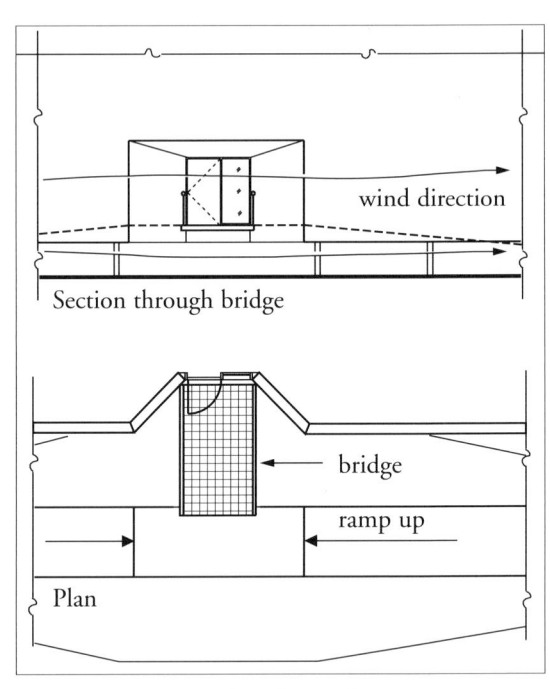

Figure 1-21 Example of gravel ramp up to an entrance bridge at the side of a raised building.

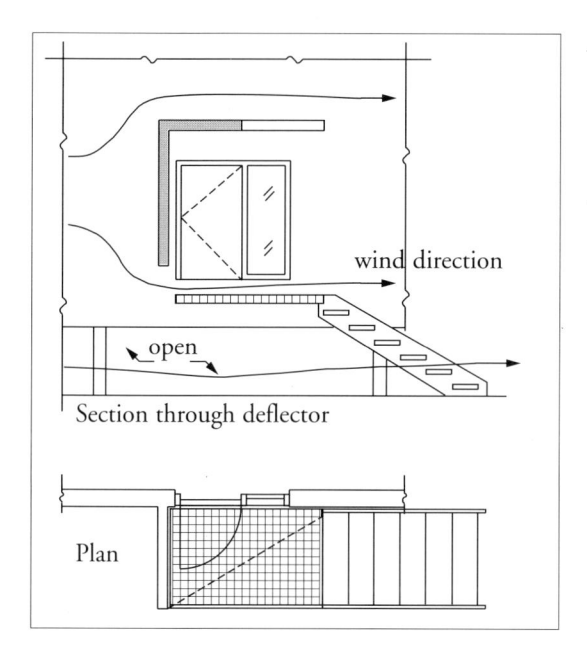

Figure 2-21 Example of wind deflector used to scour an entrance landing at the side of a raised building.

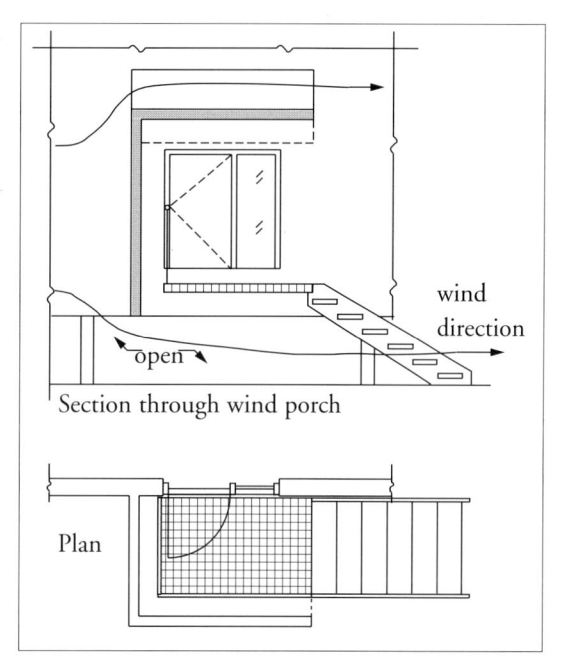

Figure 3-21 Example of "floorless" wind porch at the entrance to a raised building.

beyond the tree line. Wind blowing over the top of the building decelerates enough in the lee to drop a portion of its snow load. Either the entry side of the building or the plane of the entry itself must be oriented parallel to the prevailing wind direction. This allows the air stream to scour the entranceway instead of burying it in snow.

Deflectors: If the entrance cannot be placed to take advantage of the scouring effect of the prevailing wind, deflector panels can be installed to improve wind scour on steps and landings. A deflector, with its top edge about a metre above the roof line, interrupts the air flow coming over the roof at the lee of the building and directs it straight downward to the landing with little reduction in velocity (Figure 3-18 [8], page 96). At high wind speeds the snow particles remain in the air stream.

For an entrance in a side of the building that lies parallel to the direction of the wind a deflector projecting perpendicularly from the wall will direct air flow to scour the landing (Figure 2-21). Placed upwind of the entrance with its top edge at door height and its bottom edge 300 millimetres above the landing, the deflector forces a portion of the air flow through the gap at the bottom just above the surface of the landing. Air speed through the gap increases enough to scour the landing. The deflector also serves as a wind break for the person entering or leaving.

A wind porch will work under two conditions: its floor (usually a steel grille) must be open to the ground, allowing air and snow to pass through freely; its outer entry side must face downwind and remain fully open (the porch should not have an outer door: see Figure 3-21).

Mud, dust, and snow: Given the severity and duration of high latitude winter, mud and snow control in public buildings is often surprisingly lax. The midlatitude pattern of mud grilles in the vestibule, sometimes drained and sometimes not, prevails. These mud grilles fail primarily through low capacity and lack of regular maintenance.

Dirt and snow clinging to outerwear should be disposed of before entering the building, rather than after. Entry stair treads and landings made of galvanized steel grilles do not accumulate dirt that otherwise would be tracked inside or recirculated by wind currents around the entrance.

A soaker mat (without drain) placed on the floor of a heated vestibule adequately contains the meltwater from any snow tracked inside. The meltwater soon evaporates. Floor finish materials should retain their anti-slip properties when wet.

Meltwater from the roof that drips onto walkways near entrances quickly refreezes into a slick surface that makes sure footing impossible. Water from any source must be directed away from walkways.

Air lock: An air lock stabilizes interior comfort level by minimizing air and vapour exchange with the outside. It can be made with two doors in succession separated by a vestibule, or by one door followed by a vertical separation of space. The ideal vertical separation has the entrance door at ground level and the lobby one level higher, up a flight of stairs; the building envelope traps warm air at the higher

Storage not considered. Rankin Inlet—March

level, blocking the influx of heavier cold air from the entry door below.

Vestibule air locks are common. To work properly only the outer or the inner door can be open at the same instant: a three metre lateral separation between the two normally assures that one door remains closed while the other is opened. Some users, tired of opening two sets of doors, prop the inner doors open for long periods, negating the air lock principle altogether. The main air and vapour seal should occur at the weatherstripping of the inner door, so that the vestibule can behave like a cavity with a single opening, the outer door.

Heat: Some air locks are heated, others not. The pattern is unclear. Unheated air locks are common in West Greenland, where the winters are milder than in northern Canada. The inner and outer doors share the stress of thermal and vapour shock. The inner door separates the warm interior from the moderately cold vestibule (which is "moderated" by heat loss through the inner door); the outer door has only to separate the moderately cold vestibule from the colder exterior. No single door has to act as barrier between the extremely warm and the extremely cold. The vestibule also serves as storage for outdoor equipment such as rifles and fur garments that must be kept cold. In Canadian public buildings unit heaters recessed in a wall or ceiling pump hot air noisily into vestibules at thermostatically controlled intervals. The hot box environment shocks the senses, while the outer door, severely stressed, loses much heat energy to the outside.

The heat-no heat question can be resolved only if the program for the series of spaces called "entrance" has been properly defined.

Entrance room: Just inside the entrance, social space may be necessary for a pause, whether to wait for stragglers or for the taxi to arrive, to remove boots and overcoat, to get one's bearings, or just to talk on a chance encounter, to say hello or goodbye. Moving between outside and inside takes more time at high latitudes than elsewhere. It takes time to stamp snow off boots and clothing, time to remove outerwear, time to stow gear. The delay permits social exchange. Where people wait at the elevator, at the ticket booth, at the meeting room entrance, or at the coffee concession, an extra space allowance must be made.

Temporary storage: Where do boots and outer clothing discarded at the entrance come to rest? Too often clothing and boots end on the floor in a knee-high jumble that blocks convenient entry, and speedy exit in case of fire. Some users prefer to leave garments in a semi-cold place; others enter the building proper before removing clothing and boots. Either way the transactions are large and frequent enough to warrant study prior to design of the entry.

Orientation: Walking through a building users should not have to unroll a ball of string to be certain of finding their way out again. Make the answer to the question "Where to next?" obvious—the space just past the entrance can contain the spatial cues that suggest alternative paths: rising ceiling, open stair, sight through rooms, corridors, directory. The nature of the building decrees the number of visual cues: a community centre may be open to everyone whereas a courthouse limits public access. Above all, avoid the kind of symmetry—two or more directions to take that seem equally valid—that confuses the visitor.

Seasonal round: Typically an entry is designed for the severest effects of the winter weather. What about summer? Is all this precaution necessary then? Consider making the main entrance adaptable to the season—less onerous to use in summer—perhaps by collapsing the inner wall of the vestibule into a pocket, or by turning the vestibule into a sun room.

Door swings: Since a door swinging outward may be blocked by snow, designers often make doors swing inward, against the flow of emergency evacuation. Just as often designers forget the snow problem—exit doors swinging outward become blocked in winter, or they come off their hinges when forced repeatedly against the snow by users.

22 FOUNDATIONS

TRADITION

Native architecture at high latitudes makes small demands on the capacity of the soil to support structures. The structures are light and conduct their loads evenly to the ground. If the ground surface moves, the structures flex to absorb the movement. The snow house dome slumps to fit the new situation without disintegrating. The heat inside, from oil lamps and bodies, does not thaw the frozen foundation to the point of instability until the spring. No problem.

For a log house resting on a mud sill, a continuous strip of timber set in the ground under the walls, there is a foundation problem. The moisture in the soil freezes every fall, heaving the foundation upward. The spring thaw sees some of this movement reversed, but not in the same amounts at all points. Differential settlement of the foundation causes the structure to rack. Cracks open between the logs, and air breezes through them.

PROBLEM

Modern foundations are designed to be stable. At high latitudes permafrost conditions require extra care in their design and installation. If the stability of the permafrost is maintained, stable foundations can result. But as soon as humans disturb the ground surface—with wheeled vehicles, or a new building, a source of heat, a source of shade—the thermal regime in the soil below changes. In soils containing excess ice the change may be enough to degrade the loadbearing characteristics of the ground.

Preparing new foundation for Gertie's. Dawson—October

Differential settlement in the foundation results. The building structure racks and the building envelope splits along visible and concealed cracks open to air currents.

At the turn of the century, the citizens of the new town of Dawson, Yukon, were not preoccupied with the stability of foundations for buildings. Time was short: buildings were needed immediately, not six months later. Many of those original buildings collapsed under their own weight. The ones that remain, still leaning, still hanging on to dignity and bolstered by notions of the touristic picturesque, have a new lease on life. With its gold and gold-seeking population long departed, Dawson now labours to stabilize the foundations of its being. Slowly, expensively, the cream of the leaning buildings are being raised, squared, plumbed, and resettled on real replacements for imaginary foundations.

Clever people still make foundation mistakes. The lessons of the first half of the twentieth century were absorbed very slowly. In the mid-1960s the townspeople of Igarka in the Soviet Union were stuffing blocks of frozen carbon dioxide (-78°C) into the pile foundations below their buildings in an attempt to arrest the thermal degradation of the permafrost caused by heat loss from the spaces above. In northern Canada a few examples of mistakes, pale copies of the many that preceded, can still be seen. Although foundation engineering is an ancient discipline, foundation design in permafrost, having

Wood crib on wood footing. Taloyoak—February

begun in the 1940s, is just emerging from adolescence. At present, most foundation mishaps arise from not putting established principles into practice.

Some technical terms:

Permafrost is any soil or rock that has a temperature lower than zero degrees Celsius for more than a year.

Ice-rich permafrost is any soil or rock that contains sufficient excess ice to cause significant settlement under load once it has been thawed.

Thaw-stable permafrost is any soil or rock that contains no excess ice and that will not settle under load once it has been thawed.

Frost-susceptible soil is any soil that is susceptible to *frost heave* as it is frozen. (In general, soils that contain a significant proportion of silt are highly frost-susceptible. Well-drained, clean sands and gravels generally are not frost-susceptible and do not heave when frozen.)

The *active layer* is the surface layer of soil or rock, which thaws every summer.

The *continuous permafrost zone* is that area in which permafrost is almost always present below the ground surface, except where permanent bodies of water such as lakes and rivers are present. In Canada the southern limit of the continuous permafrost zone corresponds approximately to the tree line.

The *discontinuous permafrost zone* is that area in which permafrost is sporadically distributed, even in locations where bodies of deep water are not present.

Permafrost that consists of dry, exposed bedrock and deep, well-drained gravels (non-frost-susceptible soils) does not pose special problems for the foundation engineer. Permafrost that consists of ice-rich sands, silts, and clays does cause special problems. Chief among these are thaw settlement, frost heave, and creep.

A building founded on ice-rich soil below the permafrost table will remain stable as long as the temperature of the loadbearing stratum does not rise above freezing. If heat from the building, from utility conduits, from reflected solar radiation, from the unchecked flow of ground and surface water, even from disturbance of the adjacent ground surface, enters the ground, the upper surface of the permafrost will thaw, radically altering the capacity of that soil to bear the weight of the building. As the permafrost table sinks, so will the building begin to sink. But it will sink unevenly, given that subsoil conditions and point loads vary within the foundation footprint. This differential settlement causes the building structure to rack, and its envelope to rupture. The risk of thaw settlement increases where the permafrost temperatures are just below freezing, always the case in the discontinuous permafrost zone.

In polar regions, where near-surface soils are often frost susceptible, frost heave is the most common foundation problem. During the summer and early fall the ground thaws to a depth of one to two metres, more or less, depending on the type of soil and the microclimate at the ground surface. By late fall the active layer has started to refreeze from the ground surface downward. Ice lenses develop at the descending frost front, which causes the ground surface to heave. If the now frozen soil at the top of the active layer happens to be bonded to a foundation pile or footing, an uplift force will be exerted on the foundation as the ground surface heaves. If that uplift force exceeds the uplift resistance of the foundation, the foundation will also heave (Figure 1-22). Frost heave forces are often great enough to raise a two-storey building. In a single season the amount of uplift can range up to 100 millimetres. The foundation resettles differentially when the active layer

FOUNDATIONS 115

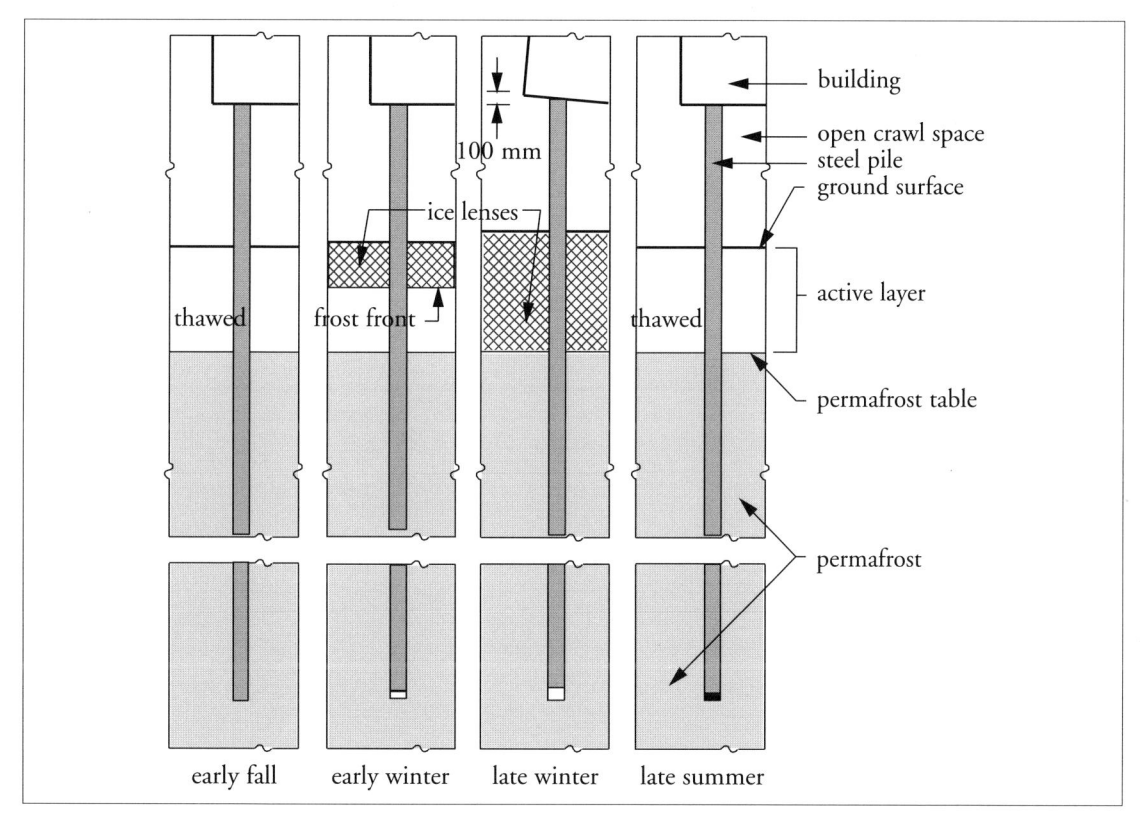

Figure 1-22 Frost heave as it may affect a steel pile to which the freezing active layer has bonded. Early fall: the active layer is fully thawed. Early winter: the active layer freezes from the surface downward, thin ice lenses form at the descending frost front, expanding the soil, heaving the ground surface slightly. If the near-surface frozen layer is bonded to the pile, uplift forces will act on the pile. If uplift forces exceed the pile's uplift resistance, the pile begins to rise. Late winter: freezing of the active layer is now complete, thicker ice lenses having formed in the lower portion; the ground surface and the pile have risen to the maximum. Late summer: as the active layer thaws again it releases the pile, which settles back imperfectly into a hole that may now be full of debris. The ground surface also settles.

Radiator portion of thermopile. Cambridge Bay—March

thaws again the following summer. Ordinary buildings cannot resist such cyclical movements for long.

Ice and ice-rich permafrost creep under load. That is, when a constant load is applied, these materials deform over time. The rate at which they creep or deform depends on a number of variables, principally the magnitude of the applied load and the temperature of the permafrost. To limit the amount of creep or settlement of a foundation occurring over the lifetime of a structure, the applied design load must be reduced sharply. Typically, to restrict creep to amounts that a structure can tolerate, the long-term design loads must be reduced to about one tenth of the allowable short-term loads.

CONCEPTS

From site to site the capacity of the earth's mantle to bear gravity loads varies greatly. We know this from

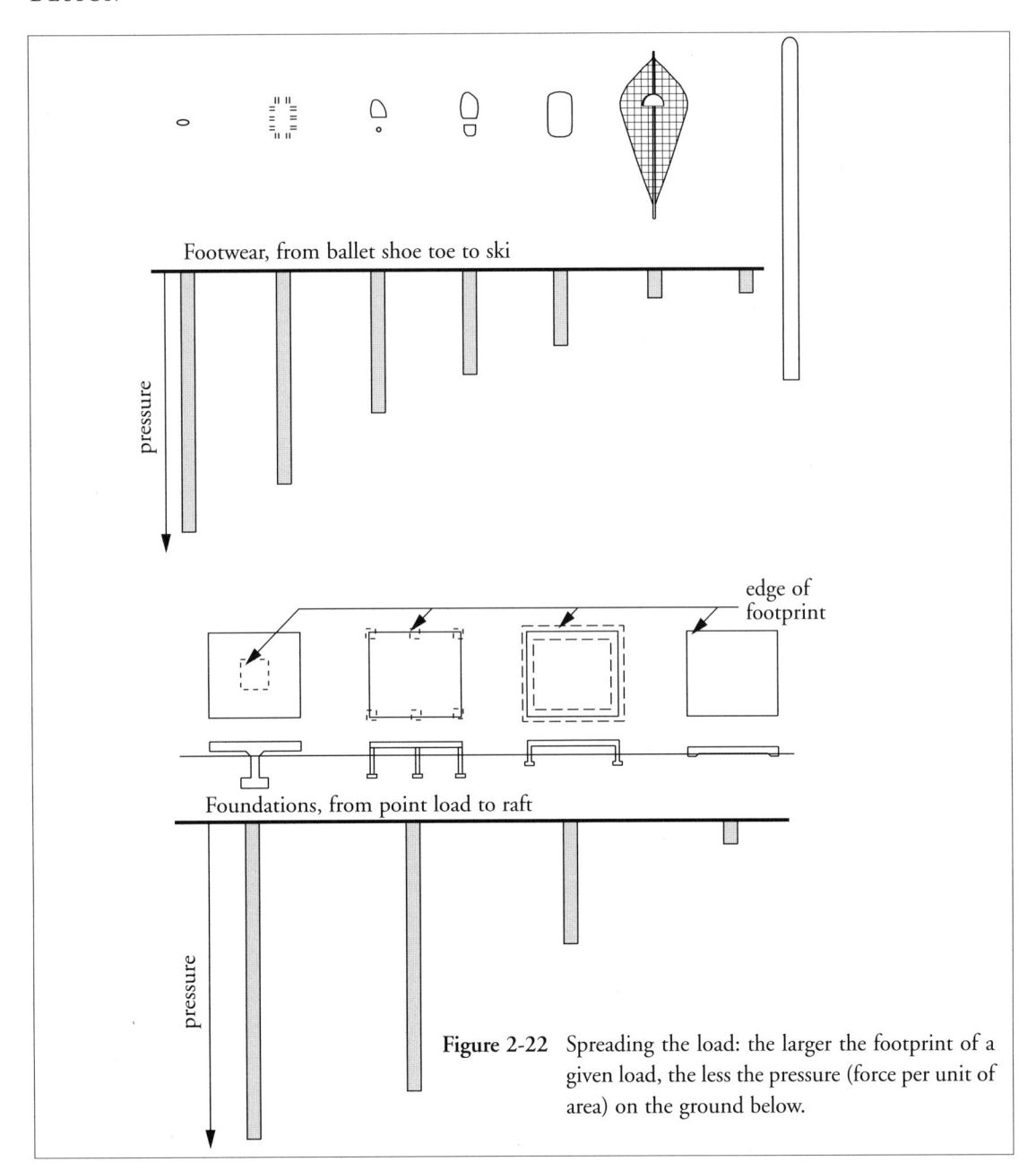

Figure 2-22 Spreading the load: the larger the footprint of a given load, the less the pressure (force per unit of area) on the ground below.

walking off the village street onto the tundra: asphalt and concrete, dust and sand, snow and ice, peat and gravel, all react differently to the load of a footstep. Some footprints are deep and some are invisible. We know that the same materials react differently again in the presence of water. Thoroughly wet or thoroughly dry, beach sand squishes away underfoot; but sand that is merely moist seems as hard as concrete. It is the same for the footprints of buildings. Figure 2-22 shows the effect on a loamy soil of any load, a human being and a small floor slab, for example, when the area of the footprint bearing on the soil changes. The smaller the footprint of a given load, the greater the pressure (force per unit of area) on the soil below, and the deeper the impression; the larger the footprint, the shallower the impression made in the soil. Spreading the load minimizes sinking. Ideally, buildings should not sink at all, so, given the soil bearing capacity, it is necessary to spread the load of the building such that any point in contact with the soil is exerting a downward pressure that minimizes settlement.

Complications occur when the bearing soil contains water, and freezing temperatures prevail all or part of the year. Unlike most other materials, water expands as it freezes, and may fill the open spaces between individual grains of soil to the point of forcing the soil to expand as well. The cycle of freezing, heaving, thawing, and settling in the active layer above the permafrost table shares certain characteristics with wave action in the ocean (Figure 3-22). A building bearing directly on the "choppy" active layer had better be as rigid as a boat, or it will break up. But a boat, no matter how big or how rigid, will toss and turn on the waves. For permanent construction it is simpler and cheaper to build a stable foundation

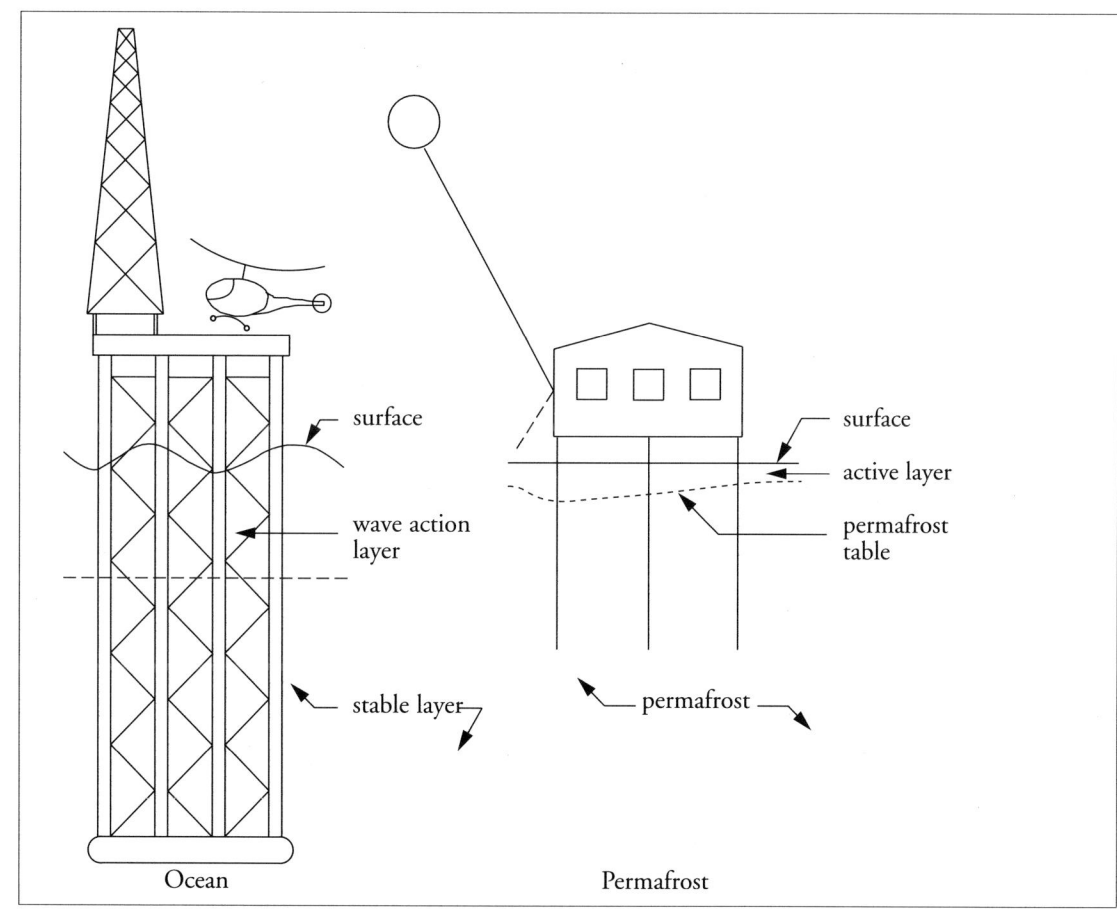

Figure 3-22 Foundations—seeking stability below the unstable layer.

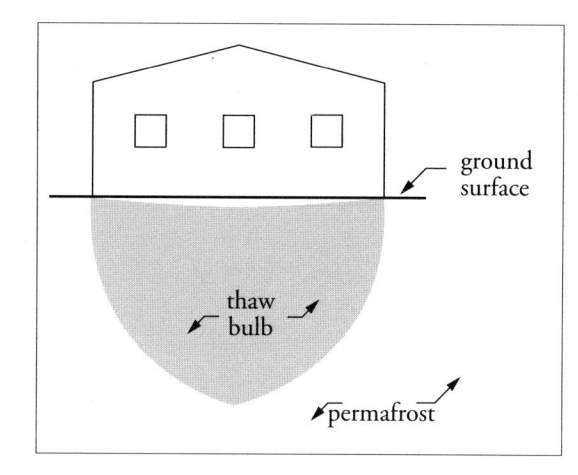

Figure 4-22 A heated building set directly on the ground develops a zone of thawed soil—a thaw bulb—in the perennially frozen soil below it.

supporting a flexible building than an unstable foundation supporting a rigid building.

Consider the "thaw bulb" in Figure 4-22. The danger, already identified, is that the heat radiated and conducted downward by a heated building set at ground level creates a zone or "bulb" of thawed soil that eventually destabilizes the foundation originally set in ice-rich permafrost. But in a different situation, where a pile foundation is placed on *thaw-stable permafrost*, such as solid bedrock, the designer fully expects the thaw bulb to degrade the permafrost. Since the foundation is on thaw-stable bedrock, no settlement occurs as the thaw bulb expands.

Where the stability of a foundation depends on preventing the development of a thaw bulb below a building, the foundation design must include a mechanism to limit the amount of heat escaping into the ground. The simplest mechanism, insulating the ground floor and raising it well above ground level, allows the prevailing wind to remove heat radiated by the building into the open crawl space between ground and floor. Refrigerating the ground below the floor, a more expensive option, achieves the same end.

Even with the building well isolated from the ground, other sources of heat must be assessed—degree of exposure to sunlight, damage to the active layer during construction, flow of ground water, surface drainage pattern, service lines—as to their impact on the heat balance in the ground below the building.

It is essential to anchor foundations against uplift forces, and not just because buildings tend to fly away on freakish winds in places such as Pangnirtung. Foundations must be anchored to counter the effect of frost heave in the active layer.

*"Space frame" on steel piles.
Resolute—July*

Consider steel piles: they are thin and smooth for ease of downward insertion in the ground. But their shape equally favours being pushed upward out of the ground. To counter this, foundation piles are usually anchored in the stable loadbearing stratum. In addition, the surface area of the foundation member in contact with the permafrost will freeze back to the soil, creating resistance to uplift. This is *adfreeze bonding*. (Some pile foundation designs depend entirely on adfreeze bonding below the permafrost table to support the load of the building.) Above the permafrost table the surface area of the foundation member in contact with the active layer can be coated with grease to minimize bonding with the refreezing soil.

Foundation design must also deal with lateral loads, from any direction, imposed on the building by wind and earthquake. At midlatitudes, lateral forces on foundations are resisted mainly by the stiffness and mass of the soil around the excavation from ground level down. At high latitudes, when a foundation design intentionally bypasses the active layer, the soil in the active layer cannot be counted on to resist lateral forces; the resistance must be provided elsewhere in the design. Battered piles—supplementary piles inserted into the ground at an angle and welded at the top to a regular pile—accomplish this, as do cross braces above grade.

SITE SELECTION

Selecting a building site whose foundation soil is competent in both the frozen and unfrozen state remains the best means of reducing the risk of foundation mishap in permafrost zones. Unfissured bedrock, exposed or very near the surface, followed by well-drained coarse gravel, head the list of favoured ground conditions. But many sites, especially in the Mackenzie River valley, offer neither. Poorly drained, low-lying sites and ice-rich ground pockmarked with pingos, ice-wedge polygons, mud boils, sorted circles, peat bogs, *muskeg*, drunken forest, solifluction, and thermokarst holes should be avoided.

Select a site that will survive the ravages of the construction process in a way acceptable to the user. Minimum disturbance of the site is the soundest policy in permafrost zones; select a site where minimum disturbance is not just required but also economically feasible.

SITE INVESTIGATION

Engage a soils consultant with broad high latitude experience—experience is nearly everything in this field, experience not just of different conditions, ice contents and below-ground thermal regimes, but also experience in moving personnel and drilling equipment efficiently into remote areas plagued by uncertain weather and conflicting priorities.

Noting local usage in foundation design is essential to an investigation. But temper it with the certainty that soil conditions can and do vary radically from site to site in the same community. Note too that the best sites in terms of simple foundation requirements already contain buildings—many of the untaken sites have dubious soil characteristics. Make note of regional preferences in installing foundations for similar conditions: foundation contractors are few, their experience and methods conditioned by the type of equipment and supplies readily available.

The soils consultant arranges and supervises the subsurface investigations required to establish the most suitable foundation system for the proposed building. This work commonly involves drilling test holes so that the properties of the soils and rock below the site can be assessed. One or more foundation systems are recommended to the owner's structural engineer based on the subsurface conditions encountered, local experience, and the foundation loads to be imposed by the structure.

STRATEGIES

Permafrost not present: Foundation design in this case is no different than that for midlatitudes. The main problem is frost heave of the active layer; placing the base of the foundation deep in a stable stratum below the deepest annual frost penetration prevents jacking forces overcoming the building's dead load and the friction between soil and foundation member surfaces.

Permafrost with no excess ice: If site surface drainage is good and the soil or rock does not contain excess ice, freezing temperatures in the soil may pose no problem. Foundation design is no different in this situation than that for midlatitudes. Subfloors or basement walls are only insulated to reduce annual heat loss from the building.

Ice-rich permafrost: Perennially frozen ground below the active layer can be considered a stable bearing stratum for the foundation as long as the ground remains frozen for the life of the building. Elevation of the building above ground and insulation of its ground floor are the simplest means of preventing thermal degradation of the permafrost. Insulation alone will not prevent thermal degradation. In a building required to be at ground level, insulation supplemented by mechanical ventilation of the enclosed space between the subfloor and the ground, or by refrigeration of the ground, should be considered. Frost heave in the active layer must be countered by anchoring the foundation below the permafrost table, by mechanical means and/or adfreeze bonding, and by reducing friction between the soil of the active layer and the surface of foundation members. Frost heave can also be controlled by placing a layer of non-frost-susceptible soil (coarse gravel, for example) over the original ground surface: this modifies the thermal regime below ground so that the active layer is repositioned inside the non-frost-susceptible gravel layer, which will not heave as it freezes each fall.

Ice-rich permafrost underlain by thaw-stable soil or rock: It costs money to raise a building above ground. If there is a thaw-stable stratum (bedrock, most likely) below the ice-rich overburden and it is not too deep to be reached economically by a pile foundation (say within 3 metres), then the building can be founded on the thaw-stable rock, and an open crawl space below the floor is not required. The building's thaw bulb permanently degrades the active layer, so that frost heave is eliminated.

Discontinuous ice-rich permafrost: Discontinuous ice-rich permafrost, its mean temperature very close to thaw, is especially fragile. For permafrost to be considered stable enough to bear loads its temperature should remain below about -3°C, depending on the soil type. The temperatures of permafrost in the discontinuous permafrost zone are usually higher than this limit. Available options include: avoiding the discontinuous permafrost altogether when selecting a site; thermally degrading the permafrost; removing the frozen soil from the site.

FOUNDATION TYPES

There are many foundation types and countless variations in use at high latitudes. Many buildings are one-of-a-kind designs (houses and garages being the main exceptions) and so are their foundations. The types most common in northern Canada are described briefly in the Appendix. They are: wood piles; steel pipe piles; gravel pads; ducted gravel pads; refrigeration; buried footings.

PITFALLS

Some common pitfalls in foundation design and installation are listed below.

Season: Since the timing of construction at high latitudes is crucial it follows that the foundation must be built quickly to permit raising and closing the superstructure before the onset of winter. Placing steel piles efficiently, for instance, means working during the preceding winter and spring months while the ground surface is frozen. Piling through a thawed active layer is usually difficult: the thawed soil may slough into the drill hole; men and machines churn the ground surface to mud. An unacceptable slowdown may result.

Inadequate supervision: A qualified person must monitor the installation of pile foundations to ensure that the piles are placed and backfilled as required, that soil conditions encountered match the predictions of the soils consultant, and that an installation record is kept of individual piles.

Unloaded piles: If the piles' resistance to frost heave in the active layer depends in part on the dead load of the building, and if the piles are installed in the spring but construction of the superstructure is deferred a year or two for any reason, the building's dead load will not be available to prevent uplift when the active layer refreezes. The piles may shift and begin to resemble a drunken forest.

High salinity: Some large construction projects have ground to a halt because soils tests failed to detect

Setting steel pipe piles. Iqaluit—April

the presence of saline pore water (from the now-receded ocean) within the permafrost. Compared to non-saline permafrost, permafrost with brine-infested ground ice supports smaller loads. The structural engineer might determine that twice the usual number of piles would have to be placed to carry the same load.

High winds: In some communities occasional high winds have been sufficiently strong to shift buildings off their foundations. Anchor buildings and their foundations securely to the ground.

Vibrating piles: Even if a pile foundation under an elevated building carries its design loads properly, the effect of wind in twisting the building laterally may have been underestimated. The building vibrates and creaks, often discomfiting the occupants. Apply adequate bracing to foundation components.

Frozen backfill: Natural gravel or sand used as loadbearing material below a new building whose thaw bulb is permitted by design to penetrate the permafrost table (because there is a thaw-stable bearing layer below) must not contain excess ice. At high latitudes, available backfill supplies may be completely frost-free only for a few weeks of the year, a period probably not coincidental with the backfilling operation. If the gravel is friable and free-moving and looks ice-free, is it ice-free in fact? Not necessarily. The unwelcome proof arrives when the contractor heats the building for the first time; the heat from the ground floor thaws residual ice in the material, the material settles, and the floor slab collapses.

Frost heave: Even if pile foundation members behave properly, the skirting detail around the perimeter of the foundation may not. The skirting, a frame supporting wire mesh that lets wind but not airborne rubbish into the crawl space, may actually touch the ground in places. As the ground heaves in late fall it exerts upward pressure on the skirting frame, which in turn raises the subfloor from which it is primarily supported. Allow at least 100 mm clearance between the bottom of the skirting and ground surface to accommodate frost heave, or design a skirting detail that accommodates vertical movement.

Cold floors: Designers who forget that warm air tends to accumulate at the ceiling, not at the floor, will have clients with cold feet. Cold floors in older houses are practically a trademark. If the occupant refuses to wear slippers he may decide to enclose the open crawl space with solid skirting or blocks of snow, to reduce the cold draft. This immediately increases heat transfer to the permafrost with predictable consequences. Design floors to feel warm.

Heated crawl spaces: It costs money to heat crawl spaces; it follows that crawl spaces should be as small as possible. In the tug-of-war between competing priorities during the design phase of a building project the size of the crawl space may grow unchecked until it becomes a walk space.

Dynamiting: How will the flight of shot rock be dampened? The usual equipment does not exist in remote communities. The contractor may opt for placing an overburden of fill over the rock to be blasted. He probably will not remove this fill from the site afterward of his own accord. Be wary of the blasting process: it may completely disfigure the site outside the building footprint.

Freeze-back time: Many mishaps occur when piles that depend on adfreeze strength are loaded prematurely by a builder in a hurry. If timing is critical, bury temperature probes at different depths next to some representative piles to monitor pile freeze-back.

Mixing foundation types: Different foundation types react differently to the same soil conditions. A house founded on a space frame will not settle or heave as much as an attached porch or outside stair supported by wood pads and wedges. In other words, the porch and stair will heave more than the house proper, creating leveling problems every year, and damage to the connections between the house and its minor structures. Place all the elements of a building on a single foundation type.

Compaction: The equipment necessary to compact backfill properly does not exist in most remote communities, and where it does exist, local interests may prevent the project from borrowing it. So compaction at high latitudes, with the exception of airstrip and highway projects that are supervised full-time, is often far from perfect. Exercise caution where good compaction is essential.

Water: The high specific heat of water and its readiness to flow just about anywhere make it a concern where there is a delicate thermal regime below ground. Water delivery drivers regularly spill water in small amounts that freeze cumulatively on the ground at filling points beside buildings. Rainwater exiting spouts at the roof line either erodes the ground or gravel pad or enters the ground below or beside the building and degrades the permafrost. Co-ordinate roof and water-delivery design with the foundation work.

Shortcuts: Consider that site decisions made in the heat of the moment may not be as good as those made as a result of careful study during the design stage. (Sometimes, of course, they are much better.) Perhaps a piling contractor, running short of full-length wood piles, recommends that some piles be installed in two pieces, an upper and a lower, to get around the problem. Sounds pragmatic. But how does the anchoring function of the lower half of such a pile work when the upper half is free to float upward under the influence of frost heave in the active layer?

PERFORMANCE

Where there is uncertainty with respect to foundation capacities, short-term load testing of selected piles before or during main construction will provide an indication of the long-term carrying capacity of a pile foundation. (The load testing process will also permit a preview of the contractor's main performance.) Burial of temperature probes with the foundation is a cheap way of providing valuable information should a future investigation become necessary in the wake of a failure. The investigator simply connects the probe leads to a portable black box and records the temperature profile.

WALLS 23

TRADITION

At the end of a day's walk, within or beyond the tree line, finding shelter from the wind becomes the first priority. The wind steals body heat; it drives rain and snow into eyes and ears. The lee of a stand of spruce or an outcrop offers basic shelter. But it is nothing compared to the walls of a tent or snow house.

The snow house wall is both windbreak and blanket. Containing large amounts of still air, the snow in the wall insulates the occupants against the cold air outside.

The wall of a tent also acts as windbreak and blanket. Made of animal pelts, it absorbs radiation from the sun and reradiates some of it to the interior, an advantage between seasons but a nuisance in midsummer.

PROBLEM

Control of air leakage heads the list of wall design objectives. Air lost through the wall has to be replaced by air coming into the building somewhere else, and that air has to be heated and perhaps humidified at a premium. Worse, the lost air carries water vapour into the wall assembly, where it may condense and freeze on concealed materials and accelerate the process of wood decay or metal corrosion. Incoming air brings rain and snow into the wall assembly.

At high latitudes the risk of trapped vapour, accelerated decay, and loss of thermal resistance is even greater. Insulation contaminated by water or ice crystals loses thermal resistance. The weight of water and ice causes the material to sag or settle. Sometimes a northern Canadian installs vapour impermeable material at the outside face of the exterior wall, in the hope of reducing drafts inside the building. This only guarantees that water vapour condenses inside the wall assembly and the building as a whole deteriorates much sooner.

STRATEGIES

The wall designer must find ways to resist air and vapour flow through the wall assembly. Having established resistance, any water vapour or water that does manage to penetrate the assembly from any direction must be permitted to escape to the outside. There should be only one vapour barrier in the wall assembly, not two or more, which together act as vapour traps.

Air barrier: An air barrier system is not an air barrier system unless it is continuous and structurally supported for 100 percent of its area. Membrane material with good airflow resistance does not properly function as an air barrier until so supported. Simply wrapping a building with such material has little effect in the long term; the material will disintegrate under the constant flexing imposed by the wind. Continuity at wall/roof and wall/floor junctions deserves special attention from both designers and builders.

Air can pass through and around many building materials. Masonry stops bullets but not air molecules; in both mortar and masonry units there are too many visible and invisible cracks caused by shrinkage and settlement. Poured-in-place concrete works as an air barrier system if shrinkage and settlement cracks can be avoided. Plasterboard works as an air barrier system if panel joints and joints at changes of plane can be kept sealed. Plywood sheathing works provided that panel joints are continuously supported.

The commonest air barrier material, spun olefin fibre sheet, while fairly impervious to air and pervious to vapour, acts as an air barrier system only when fully supported on both sides by materials more resistant to airflow. In practice this means placing the membrane inside a sandwich of plywood or plasterboard. Too often builders install this material with support on one side only, and sometimes without continuous support on either side. The sheet bulges first in one direction then in the other as gusts of wind beat around the building. It tears at its moorings. Some builders even cut the sheet into small pieces to make installation on the wall easier. Such lapses of understanding mean lost heat energy, wasted material, and increased potential for rot or corrosion inside the wall.

The precise placement of air barrier systems in the wall assembly cross section deserves careful study. If the material selected does double duty by being impervious to vapour as well as to air then it must be placed on the warm side of the insulation, the only possible position for a vapour barrier at high latitudes. If vapour-permeable, the air barrier system can be placed next to the vapour barrier or somewhere closer to the outside, provided structural support is available. At the outside position, the air barrier system permits the escape of vapour traces to the outside and does extra service by reducing the air movement in and out of the wall insulation caused by wind or by convection inside the wall assembly. (In a heated building the temperature differential between the top and the bottom of an exterior wall creates convection cells inside the wall assembly: air rises slowly up the warm, inner face of the wall and descends along the cold, outer face.) Placed near the outside of the wall, outside the structural components, the air barrier system's chances of being installed without breaks in continuity are excellent. At this location the air barrier system must tolerate wide swings in temperature.

At midlatitudes some building experts promote plasterboard (drywall) as an air barrier system. Plasterboard performs adequately under well-controlled conditions of construction and maintenance: all joints and junctions in the installation must remain sealed for the life of the building. At high latitudes such assurances cannot be given. Since settlement and shrinkage cracks must be anticipated, the most careful installation of a plasterboard air barrier system may fail before the building is commissioned.

Vapour barrier: This barrier exists primarily to prevent passage of vapour by molecular diffusion. The air barrier system elsewhere in the assembly restricts the movement of air/vapour currents from one side of the envelope to the other. (The vapour barrier may also serve to limit convective air circulation between the heated room and the wall cavities.) The vapour barrier may consist of polyethylene film, aluminum foil backing on plasterboard, or two coats of oil-based paint, for example. The service temperature of the vapour barrier must remain above the dew point temperature of the air space enclosed by the wall, to prevent surface condensation. At high lati-

Balloon frame. Uummannaq—August

tudes, this puts the vapour barrier on the warm side of the wall insulation. The barrier should be continuous and relatively puncture-free. (The amount of vapour diffusing through a small opening in the vapour barrier is negligible.) A building under construction should not be heated until the vapour barrier is in place and inspected for continuity. Failure to observe this cardinal rule leads to condensation of water vapour inside the insulation, where it will remain at least for the first winter and perhaps longer, reducing the value of the insulation while promoting rot in wood members.

Thermal barrier: Building insulation was invented to slow the rate of heat transfer from a heated enclosure to a surrounding heat sink—the great outdoors in winter. In wood frame construction glass fibre insulation batts remain the most convenient insulation to install because they fit snugly between the studs in space that would otherwise be unused.

Compared to the insulated space between the studs, the studs themselves act as thermal bridges that reduce the total thermal resistance of the wall assembly. An outer layer of insulation across the studs greatly reduces this bridging effect. (Even a thin layer of insulation across the *inside* face of the studs has the detrimental effect of lowering the temperature of the vapour barrier at stud locations to near or below the dew point temperature of the inside air, when the outside air temperature drops steeply below -20°C.)

Rain barrier: All the attention focused on keeping water vapour inside the building must be balanced by concern for keeping snow and rain out of the building. The air barrier system does most of this work: wind-driven rain cannot penetrate very far into a cavity that has no exit. This is the basis of the "rain screen" principle. The outermost layer of the wall, the rain barrier, catches the brunt of the foul weather; a vertical air space separates the rain barrier

*Platform frame.
Rankin Inlet—October*

from the rest of the wall assembly. (Installed somewhere between that air space and the vapour barrier, a different wall component works as the air barrier system, preventing the communication of rapid air pressure changes to the enclosed space.) Small perforations at the bottom of the rain barrier ensure that air pressure remains equal on both sides of the outer layer. With air pressure equalized there is no force available to drive rain or snow into the air space behind the rain barrier.

With the assurance that rain and snow will not penetrate the rain screen, it remains to select materials and finishes for the rain barrier that shed water, dry rapidly after wetting, and tolerate freeze-thaw cycles.

WALL ASSEMBLY TYPES

Platform frame: The *platform framing* method dominates wood construction at high latitudes in Canada. Simple and economical, it has outlasted countless prefabricated construction schemes. It is quick. It uses interchangeable parts. It is familiar to local builders and carpenters. Load-bearing structure and infill panel are one and the same.

The contractor builds the floor structure up to the level of the subfloor, creating a platform for all subsequent framing work. Wood framing members for a section of wall are assembled flat on the subfloor. The sill plate is toe-nailed on edge along the side of the building where the panel will eventually stand. The builders complete the panel as far as the exterior rigid insulation. The nails through the sill plate serve as hinges when the crew tilts the panel into an upright position. As each wall panel takes its place in line, shelter from the wind is increased. Outside scaffolding will only be necessary for the installation of cladding.

Balloon frame: A builder using the *balloon framing* method erects the wall frame first, setting it directly on the foundation. The floor joists are then fixed laterally to the wall studs. The improvements in wall continuity around floor elements afforded by this method seem slight compared to the absence of a working platform for the erection of walls and roof. Although little used in northern Canada, balloon framing methods are current in Greenland.

Post and beam: In northern Canada large buildings that take two building seasons to complete are often framed in steel—columns, beams, and open-web joists. The envelope, however, often consists of wood frame construction.

Steel portal frame: In northern Canada light industrial buildings used as arenas, garages, firehalls, and workshops usually consist of a steel portal frame with steel sheet lining and cladding. Structural members are manufactured in southern Canada. Steel buildings designed by the manufacturer often fail to incorporate details that permit air and vapour barrier continuity inside the exterior wall, with predictable results. Designing steel sheet walls with an air barrier system adequate for high latitude climatic conditions requires special study. For instance, trying to form an air barrier system in a wall that is penetrated at regular intervals by main structural elements defeats the purpose of the exercise.

EXAMPLE

The example in Figure 1-23 shows how various components work together in a platform frame wall:

Plasterboard: Gypsum plasterboard provides fire resistance, a base for paint finish, high density for sound attenuation, and high specific heat for thermal mass.

Polyethylene vapour barrier: The membrane is placed on the warm side of the insulation. No space

Inside | Outside

RSI
0.12 inside surface (air film)
0.08 gypsum wallboard
0.00 polyethylene vapour barrier
3.50 fibre insulation (wood studs)
0.10 plywood sheathing
0.01 spun olefin membrane ⎤
1.32 polystyrene insulation ⎬ air barrier
0.10 air space (wood strapping) ⎦
0.18 prestained wood siding
0.03 outside surface (air film)

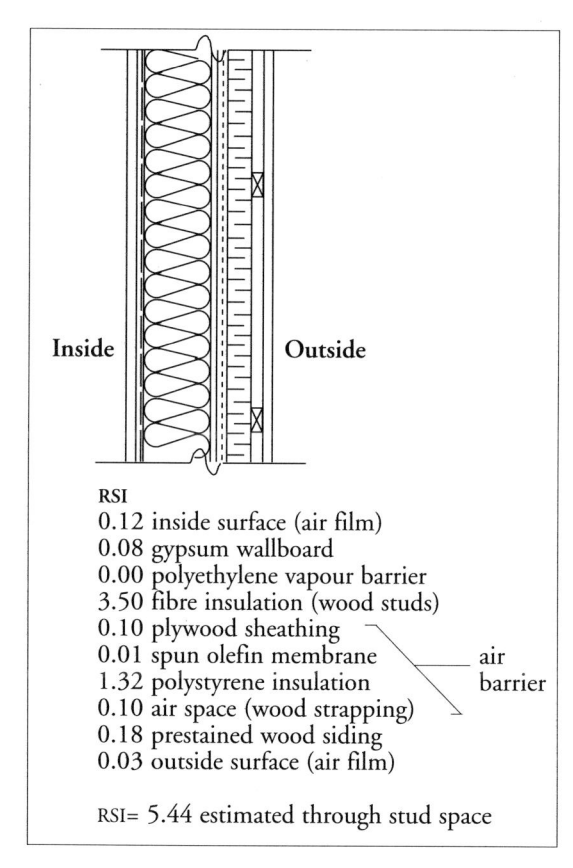

RSI = 5.44 estimated through stud space

Figure 1-23 Example of a high latitude exterior wall assembly.

allowed for cable runs and electrical devices; use supported *polypan* sealed to polyethylene film where installation of an electrical device cannot be avoided.

Glass fibre insulation: Friction-fit batts fill the space between the loadbearing wood studs.

Plywood: Sheathing provides diaphragm support for the structure, and must be exterior grade to withstand exposure to weather during construction. Given plywood's resistance to vapour diffusion, vapour traces escaping outward pass through the joints between panels.

Olefin film: Structurally supported on the inside by the plywood sheathing, and on the outside by rigid insulation boards stiffened by wood strapping, this combination functions adequately as an air barrier system for most purposes.

Polystyrene insulation: This layer reduces the thermal bridge effect occurring at the wood studs while improving overall thermal resistance. Because extruded polystyrene is nearly impervious to vapour flow, joints between boards must be left open to permit diffusion of vapour traces to the outside.

Wood strapping: Strapping provides both a nailing strip for the wood siding and an air space for ventilation of the inboard side of the wood siding. Ventilation reduces warping and cupping of the siding boards caused by different rates of drying between inboard and outboard faces.

Horizontal corrugations. Gjoa Haven—February

Prestained wood tongue and groove: The wood is prestained at point of shipment so that winter weather does not delay finishing work. The choice of sealer rather than paint to stain the siding ensures the escape of vapour released by the wood after soaking by rain. The siding is set vertically to permit rapid drainage of rain water. Select narrow width boards (say 100 mm) so that cross grain shrinkage has the least effect on joint stability.

WINDOWS 24

*At home.
Gjoa Haven—February*

TRADITION

Inside the snow house, just above the entry tunnel, a patch of daylight pokes through the dome. A small skylight diffuses light from a pane of sheet ice wedged into the snow block structure. Minutes pass before the eyes adjust to the dimness of the interior. Finally, the patch of daylight becomes an indicator of the weather outside: bright sunlight, high cloud or low cloud, moonlight. From the outside, during the long polar night, that patch of light reverses itself, becoming a patch of lamplight distinguishable at a distance: a beacon, a sign of life, of warmth and shelter, of help if needed, of friends and family. Seen at the end of an exhausting journey a camp site without a beacon chills the heart.

PROBLEM

Whether top or bottom-hung, double-hung or sliding, windows in use at high latitudes tend to fail in winter because they are expected to double as air vents. They are replicas of windows designed for midlatitude climates. In winter the gaps between the opening sash and the window frame leak moist warm air. This moisture condenses on contact with the cold window edge and freezes the sash to the frame. The window remains closed or partially open for the rest of the winter. Neither the sash nor its hardware are designed to resist the blows of a user intent on adjusting the draft of a jammed window. Hardware breaks, casements break, panes break.

Second, windows fail because the panes may be so poorly sealed or wrongly placed in the thickness of the wall that the temperature of the inboard window glass frequently drops below the dew point temperature of the air inside the room. Placed in or near the plane of the outer face of the exterior wall the window glass benefits least from the warming and drying effects of room air circulation. Vapour condenses as ice crystals on the coldest portions (the lower edges) of the window first. As ice accumulates, blocking the view and much of the available daylight for the entire winter, the window dehumidifies the room. The presence of ice and meltwater on window materials and adjacent finishes accelerates their deterioration and shortens the useful life of the window.

Third, designers often make window openings too big in area and too long in perimeter joint (between frame and wall structure). Too much heat energy is lost through glass and leaky joints in winter, and too much heat is captured greenhouse-style in summer.

Fourth, designers are held captive by building practice. Despite countless small improvements over the years the window has not been reconsidered from scratch for several decades, much less redesigned for high latitude conditions.

STRATEGIES

Purpose: In Rankin Inlet blizzards and whiteouts spring up as readily now as they did hundreds of years ago. Losing one's way at the edge of town is a distinct possibility. One new building at the northwest edge has a prominent window that acts as a beacon. Unlike any other window in the town it consists of four acrylic domes arranged in a square opening. It lights a stairwell during the day. At night it transmits artificial light from the stairwell toward the town road with the message: "Welcome. This is the last stop before the open tundra." A window without a purpose is an opportunity lost.

Function: A window in a wall or roof works well when it provides a steady mix of view, daylight, fresh air, and contact with the outdoors. Although a multi-function window for high latitudes can be designed, the uncertainties of production, markets, installation, operation, and maintenance make the chances of real success very slight. The pragmatic designer reduces the stress on a window by separating some window functions from others. By separating the fresh air function from view and daylight functions, for instance, the window pane can be fixed in the opening and sealed against unwanted air and moisture movement. Fresh air can be let in via a small vent port in the wall above, beside, or even remote from the window.

Orientation: Designers often decide to place most windows in the walls facing the equator, to capture winter sunlight. This reduces the net heat loss through all windows, since fewer windows face the pole side of the building, a sunless exposure swept by the cold prevailing wind. In summer a blind facade facing the pole also shuts out the low-angled midnight sun that might otherwise disturb the sleep of occupants. A window facing the rising sun will help rid the room of its night chill. A window facing the sea, a rock face splashed with lichen, or a copse of stunted trees is preferable to one that looks onto the neighbour's living room window.

Position: Locating a window in the wall so that one can see out both sitting and standing may be desirable; placing it so that incoming daylight reflects off adjacent wall surfaces improves the distribution of light in the room. Consider that winter sunlight, low-angled and reddish much of the day, can reach and surprise the farthest corners of the building through judiciously placed windows.

The single most important antidote for cold window syndrome is the placement of the window in or near the plane of the interior face of the wall. In that position circulating room air, having unrestricted access to the entire glass surface, both warms and dries the window.

In winter, cold window glass cools adjacent room air enough to make it fall to the floor (unless the current is reversed by a working radiator below the sill). The resulting cold draft is uncomfortable to sit or work in. Worse still, little of the body heat radiated in the direction of the window returns to the source: there is no insulated wall to reradiate heat. On the other hand, a designer may deliberately use a cold window surface to set up or reinforce a desired pattern of air circulation in the enclosed space. The cold window effect should figure in the heating and ventilating plan of a room just as much as solar heat gain through windows.

Size: The larger the window, the greater the heat loss (winter), the greater the heat gain (spring, summer), the more the light, the less the glare, the greater the risk of breakage, the greater the expense and inconvenience of replacement. Since the thermal resistance of a triple-glazed window (RSI 0.66) is only a sixth of a wall's (RSI 3.50), the size of the window is more important to energy conservation than the number of panes.

Fixed windows and operable portholes. Yellowknife—July

Shape: A circle encloses the most area with the least perimeter, outperforming the square. Similarly, a square encloses more area with less perimeter than a rectangle. For this reason a circular window thermally outperforms a square or rectangular one, but it is impractical to buy (acrylic domes excepted), to install, and to seal. But for a given area it has the least amount of perimeter joint, and therefore causes the smallest amount of air leakage. A square window is thermally the next most resistant and a rectangular one the least.

*Windows, hoods and operable port vents.
Iqaluit—April*

Information: Informing the user as to the purpose and function of windows, particularly operable ones, will improve window performance. Even small gestures—a placard in the appropriate language saying "Open only in summer" affixed to the sash of an operable window—would make a difference.

BARRIERS
Air barrier: Continuity of wall or roof air barrier system with the window or skylight is essential. Glass itself is an air barrier, but the joints between glass and the rest of the building tend to leak air. Transfer of heated air and vapour to the outside result in heat loss and structural damage due to excessive releases of moisture inside the wall, roof, and window assemblies. Care in detail design must be followed up with vigilance during construction to ensure that the air barrier system in the wall is sealed and secured to the window frame.

Vapour barrier: Window glass is a barrier to vapour diffusion, but the joints and materials around it are much less so. The potential for vapour diffusion through these joints and materials can be reduced by placing the plane of the insulating glass in or near the plane of the wall or roof vapour barrier, so that the total area of these least resistant joints and materials is kept to a minimum. If the material between the joints and the window glass (often wood) is vapour-permeable, an oil-based paint finish on its interior surfaces will suffice to control vapour flow. To ensure continuity of the barrier at the joint between frame and wall, the entire window and casing is "bagged," prior to installation, with polyethylene film set in a bead of caulking along the hidden sides of the frame and stapled. Once the bagged window is installed in the rough opening, the portion of the bag covering the window glass is slit and folded back in four directions to lap the vapour barrier material in the wall. The lap joint is sealed with caulking.

Thermal barrier: The thermal resistance of a single window pane (RSI 0.15) is very low; the temperature of the glass in winter will certainly be below the dew point of the heated space it serves. A massive build-up of condensate will freeze to the glass, only to melt and wet the adjacent finishes as soon as the outside air temperature rises. The frost accumulates first at the coldest spots, the bottom and side edges, and last at the warmest, top centre. There are several ways to raise the service temperature of window glass in winter.

Insulating glass (the glass industry's name for double, triple, or higher glazing units) is normally substituted for the single pane window. It consists of two or more panes of glass separated by an airspace and held in that position by metal edge trim permanently sealed at the factory. The airspace, if not thicker than about 12 millimetres, acts as insulator. The thermal resistance of the double pane is RSI 0.35 and the temperature of the innermost pane rises to near or higher than the typical dew point temperature in a heated room. A triple pane (RSI 0.66) makes it a certainty under normal conditions.

Second, insulating value and glass temperature can be increased at the factory by application of a low-emissivity coating on the outer side of the inboard pane of a double sealed window. Colourless, this coating traps far infra-red radiation (radiant heat) emitted by warm surfaces inside the room, increasing the effective RSI value of a double-sealed window to the equivalent of a triple-sealed window (RSI 0.66).

Third, the window can be ventilated. The building designer ensures that room air circulation flows against the inner pane of the window, perhaps by locating heating registers below the window. Forced air outlets placed specifically to ventilate windows have been tried with mixed success due to uneven airflow over the breadth of the window. Room air movement against the glass does two things: it keeps glass temperature above dew point temperature under any but the severest outside conditions; and if condensation does occur at the coldest spots of the window, the circulating room air, being drier than any air layer stagnating against the wet glass, evaporates the condensate.

Fourth, a "storm" or outer pane of glass can be installed to distance the chilling action of the wind from the insulating glass. This reduces heat loss and permits a relatively still layer of air to form against the outboard pane of the double-glazed window (not to mention at the inner face of the storm pane). Since still air is a good insulator, the resulting thermal resistance (about RSI 0.08) is added to the value for the insulating glass. This storm pane must be installed "dry" (unsealed), so that any vapour escaping the room around the edges of the insulating glass may migrate freely to the outside. At the same time, the dry mount must be sufficiently tight to prevent wind-blown snow from entering the space between the storm pane and the double-sealed window.

WINDOW TYPES

Windows being installed in northern Canada are close relatives of windows used in other parts of North America. Adaptation to high latitude conditions and needs is negligible. The strengths and weaknesses of the common window types—awning, bay, casement, sliding sash, fixed sashless—are reviewed in the Appendix.

SKYLIGHTS

Skylights designed for use at lower latitudes fail commonly in two ways: condensation forms on inner surfaces, freezes, and then thaws on exposure to sunlight, dripping on the occupants below; they are often poorly installed through the roof assembly, resulting in the damming of meltwater on the roof upstream of the unit and eventually in leaks.

Skylight over the school's resource centre. Taloyoak—August

Upstream water or snow must be diverted from the skylight, just as an island deflects the flow of a stream, before it reaches the upper edge of the unit: this can be done with a "*cricket*" in the roof slope above the unit, or by installing a circular skylight or a square one with the unit rotated 45 degrees so that one corner points upstream.

Condensation usually occurs because the skylight is at the top of a narrow shaft (through the roof structure assembly) full of stagnant warm air: the top few centimetres of the air column in contact with the cold skylight surface cools below the dew point, however, causing condensation. Normally, window glazing and seal are placed at the warm side of the building envelope. Because the skylight does double duty as a portion of the roof at or above the roof plane itself, it is located at the cold side of the roof assembly. To counter the negative thermal effects of such placement, interior air circulation against the skylight, and a water collecting channel around the inside of the opening sufficiently large to act as an evaporating dish, must be provided.

DAYLIGHT

Little is understood in a practical way about the relation of windows to interior daylighting, although people readily agree that daylight is essential to human comfort. At high latitudes, particularly in the barren grounds, where shade trees are non-existent, most of the ambient daylight incident on vertical window glass passes through into the room. With snow on the ground nine months of the year, reflected light also contributes to the total. During the winter night the same snow cover seems to amplify the available moonlight and starlight outside the window.

Glare, the unequal distribution of light—the brightness of the window seen against the lower

Porthole at interior face of exterior wall. Yellowknife—April

average light level inside the room—poses a major problem. The smaller the window in a room, the less ambient daylight there is and the brighter, the more glaring, the window seems in contrast. To reduce glare without increasing window area, the depth of the window frame can be increased and the frame outside the glass painted with a light-coloured sealer so that more daylight is reflected indirectly into the room to supplement the direct daylight.

FRAME

The window frame must resist the wind loads imposed on the glass by the wind, and perform a seamless connection with the rest of the building fabric. It must also prevent building movement and vibration from stressing the brittle glass and the weak seal between glass and frame. The building vibrates in response to buffeting by the wind, the cycling of major appliances, and the presence of moving occupants. The building may move in response to differential settlement or heaving of the foundation, to changes in relative humidity of the air (swelling/shrinking), and to changes in ambient temperature (expansion/contraction). The window frame's thermal bridge effect can be minimized by detailing a discontinuity in the frame and/or by the selection of frame materials (such as wood) with low thermal conductivity.

MATERIALS

Sealed double panes travel to the site more safely than single panes. Unfortunately, the quality or durability of the factory seal between panes cannot be checked visually. A failed seal looks exactly like a good one. The result of a broken seal is water vapour entry between the panes, *hoarfrost* in winter, and reduced transparency year-round.

Building designers select acrylic panes occasionally, to minimize breakage by vandals. Acrylic resists scratching poorly, however, flexes too much in the wind, and changes dimensions too readily with temperature change, making it difficult to maintain a seal at the edge. Polycarbonate sheet is more scratch-resistant but comparatively expensive.

Domed acrylic shapes are more successful than flat sheets, being very rigid once secured in the frame, and are available in sealed double or triple thicknesses. The dome layers are not concentric, however. This leaves an air space at the centre too wide to prevent heat loss by convection and one at the edge too narrow to act as an insulating layer. The temperature of the edge may remain below room dew point temperature much of the winter.

The high thermal conductivity of anodized aluminum or pressed steel frames results in a thermal bridge through the wall unless thermal breaks are included. Metalwork is hard to modify or repair with local resources, and replacement is complicated by the fact that the window model or frame profile may be discontinued at the source.

Polyvinylchloride (PVC) window frames of various types and qualities have been tried over the years. Brittleness and cracking of this plastic under high latitude service conditions plus high first cost have slowed their general acceptance. The thermal resistance of hollow PVC edge details is not much better than that of metal. PVC cannot be readily modified in the field.

Milled wood still makes the best window, despite the regular maintenance required (painting, for example). Being a natural insulator (about RSI 0.23 per 25 mm thickness), dry wood reduces the thermal bridge problem appreciably; it can be modified on site; it is fairly cheap (for the present).

HARDWARE

Standard window hardware is not designed for the rigours of high latitude environments. Suitable product lines cannot be economically produced for such a small market. Designers can compensate by specifying commercial grade, avoiding models that create large thermal bridges, reducing the amount of hardware installed and, where it is installed, making it do as little work as possible. The latter can be achieved by reducing the size of the opening so that the forces of opening and closing exerted on the hardware through a shorter lever arm are diminished.

Operating handles and gears often break down in use. Simple levers or latches are superior in this respect. Weatherstripping is essential: in a vent port a double seal consisting of a closed-cell foam plastic gasket inside and resilient metal strip outside performs adequately.

OTHER FEATURES

Shades, shutters, and drapes: Shades and shutters placed near the inside face of window glass prevent circulation of room air against the glass, thereby reducing the temperature of the latter to below dew point; placed much farther away from the glass they cease to function fully as shades. The ideal position for shades and shutters is outside the window. This would improve the thermal resistance through the window and increase glass temperature. Unfortunately, such outside elements would be at the mercy of wind, snow, and vandal, and a through-the-wall mechanism (thermal bridge and air leak) would be required to operate them. Venetian blinds installed in the thin air space between the two panes of glass are overrated: they obscure the view even in the open position, do not perform as well thermally as claimed, and are subject to never-repaired breakages. If window glass is placed close to air circulation, a narrow venetian blind that can be raised out of view and repaired easily works well. Drapes installed so that they do not stifle air flow against the glass also work well.

Fly screens: Every opening window and vent port is screened against black flies and mosquitoes. This raises another objection to the opening window: to prevent entry of insects the entire window area must be screened, effectively reducing transparency by half. A fly screen in a vent port is not a visibility problem and its small size makes it less prone to tearing than a large screen.

Window hoods: The absence of projecting roof eaves in some high latitude building designs exposes windows to more rain, meltwater, and ice than is the case at lower latitudes. A window hood is worth considering. Where a hooded vent port is located above a window, its hood can be extended to protect the entire width of the glazing (Figure 1-24). Since sun angles remain low most of the year, the hood has no effect as a shade.

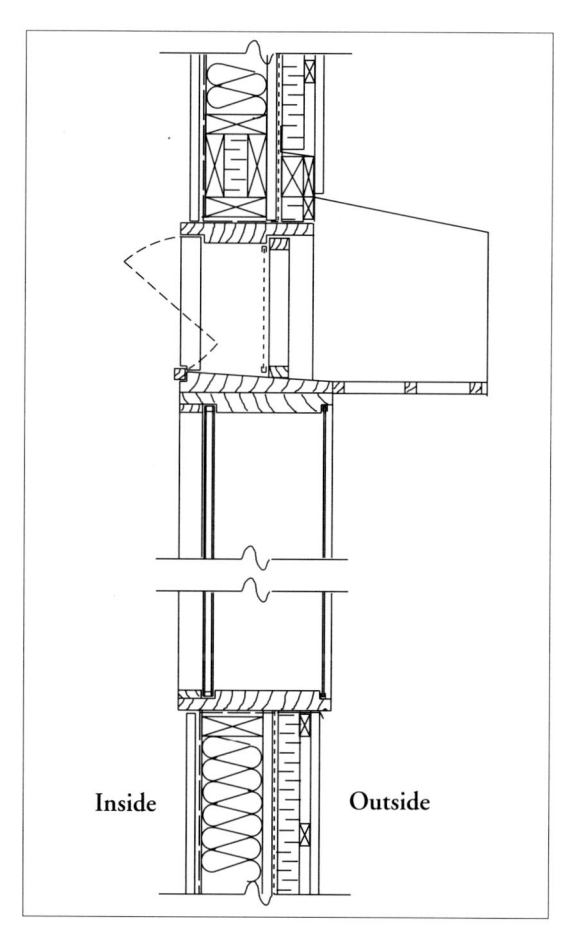

Figure 1-24 Example of a high latitude window assembly with double inner glazing (sealed), single outer glazing (unsealed), operable vent port, fly screen, and window hood.

DOORS 25

At school. Lutselk'e—March.

TRADITION
In the teeth of a blizzard, a snow house, however hastily built, seems to work a miracle. The occupant wedges a block of snow into the outer entrance of the tunnel. The tunnel protects the inner doorway from the elements. A tunnel floor lower than the sleeping platform forms a vertical air trap to prevent cold outdoor air from welling up into the living space. An inner door is not necessary.

In the summer tent a flap of caribou skin takes over from the snow block. The flap stays closed when the wind is up or the flies are out, and reopens for light, air, and a view of the tundra. Whatever the season, the material of the door and the wall is the same. Whatever the door, its threshold stays at ground level.

PROBLEM
Some designers expect a door to do the work of a modern wall when closed, resist gusty winds and casual abuse, and accept thousands of openings and closings in all kinds of weather without drawing attention to itself. It is not possible.

Exterior doors typically installed in high latitude buildings are designed and mass-produced at midlatitudes to suit temperate climatic conditions. They resist high latitude climates and usage poorly. Too much is expected of them. If opened frequently they cannot reliably separate a -30°C blizzard from a +24°C foyer with a total thickness of less than 45 millimetres. The air temperature differential causes unequal thermal stresses on opposite faces of the door: the door warps and does not close tightly against the weatherstripping. No longer truly aligned with the pin axis of its hinges, the warped door strains at its attachments. The plastic weatherstripping becomes brittle in the cold and breaks—air leaks abound. The passage of booted feet further damages the weatherstripping of the threshold, causing breakage and an influx of powder snow. Snow granules pack into door jambs: the door sticks. Water vapour condenses as frost, where inside air escapes around the edges of the door: the door jams, and the screws holding the hinges work loose under the wrenching motion imparted by the person trying to open the door. Latch hardware, flimsily manufactured, fails to resist the extra forces. It makes a thermal bridge to the outside causing condensation to freeze on the inside, jamming all movement. The hardware breaks. Replacement sets may not be in stock.

STRATEGIES
The ideal door for high latitudes has not yet made an appearance. Figure 1-25 shows that only half the surface area of a conventional door has proper support: the upper and lower corners on the latch side cannot be expected to close tightly to the door frame.

Designers and owners consider doors with multiple latches, like those in aircraft, to be too expensive to build, install, and maintain. The well-insulated doors developed for commercial walk-in freezers have been tried at high latitudes in houses and public buildings but, since typical freezer doors open outward, snow drifting against the doorway blocks their operation. Not designed to resist casual abuse by the general public, freezer doors also demand constant maintenance.

Since the common door cannot be made to match the winter environment, the environment must be modified locally to suit the capabilities of the door. The central strategy is to separate functions so that each component of an entry system faces tolerable stresses.

First, place two doors in succession in place of one, so that each does a portion of the work of separating the indoor and outdoor environments. The vestibule space and the outer door protect the inner door, the primary closure, from the brunt of the stress imposed by the outdoor environment. The inner door seals in indoor air and vapour. The outer door functions as weather break only: its imperfect seals keep out rain and snow. The vestibule environment, colder than the indoor environment, remains warmer than outdoor conditions because of heat gained through the inner door. This arrangement serves to decrease the temperature and humidity gradients across the primary closure.

Second, treat main entries and other exits differently. Entries—full-time doorways—should have the two doors separated by a vestibule, but exits, infrequently used, may be single doors, provided they give onto a circulation space that does not require a high level of comfort.

Third, separate the functions of door hardware to reduce stress on individual components: install independent pieces of hardware.

Fourth, select weatherstripping that functions at low temperatures and install it so that damage and dirt resulting from foot traffic do not diminish its performance. Select exterior doors and frames equipped with thermal breaks to reduce conductive heat loss to the outside.

BARRIERS

Compared to a wall assembly, a door and frame work poorly as a barrier to air, vapour, and heat. A door's resistance to heat transfer may be only a tenth or a fifth that of the adjacent wall. While the door leaf itself may resist the passage of air and vapour quite well, the joints between door and frame and between frame and structure resist inadequately. (Of course a door left ajar resists nothing.) Designers work to the severest winter condition, but there is nothing to stop development of a door and entry design that adapts to summer conditions. One set of doors might simply slide away into hiding for the summer months, leaving a single doorway to cope with the mild weather.

DOOR TYPES

Door types commonly used in northern Canada—steel, solid wood, plastic-faced wood, glazed, interior, overhead—are described briefly in the Appendix.

HARDWARE

At high latitudes door hardware suffers from being designed for mass markets at temperate latitudes. Hardware standards suitable for high latitudes do not exist, while availability of heavy duty hardware and replacement parts remains unreliable. The lockset on a conventional door installed at midlatitudes does four things: the door handle can be pushed or pulled to move the door leaf; the handle can be turned to open the latch; the key can be turned to lock the latch; the latch secures the door in the closed position. When a door freezes shut against its

Going home. Rankin Inlet—January

weatherstripping and the lockset, congested with hoarfrost, refuses to operate, the person inconvenienced attempts to pull the door, turn the handle to open the latch, and unlock the lockset with main force in one or two motions. The hardware, already stressed by low air temperature, breaks down, usually before the door opens. In this situation a door requires separate push and pull handles, a keyed dead lock independent of the door latch, and a levered latch set. (Replacing a slippery door knob with a lever door handle shifts greater force to the

At home. Pelly Bay—February

delicate working parts inside: internal breakdown can result if the lever handle itself does not break off first.)

Hinges: An open door has the potential of acting as a big lever operating against the door jamb with the hinges as fulcrum. But an obstruction at the bottom edge of the door, such as ice, moves the fulcrum away from the hinge position and converts the door into a lever against itself. The door comes off its hinges. Ball-bearing hinges are essential on an exterior door. Hinges must be screwed through any wood door frame into the building structure to ensure solidity. Pressed steel door frames must be reinforced at hardware locations to receive hinges. Provide exterior doors with robust anti-gust hardware to reduce wind damage.

Latchsets: The latchsets commonly available are too flimsy for use at high latitudes. Good quality door hardware pays for itself in lower maintenance costs. The door handle design must allow for operation with mittened hands—this usually means a lever handle.

Push and pull: By their shape push plates should signal "push" to the user and pull bars should indicate "pull"—at any latitude. A pull bar with a gap behind it too small for a mittened handhold should be replaced.

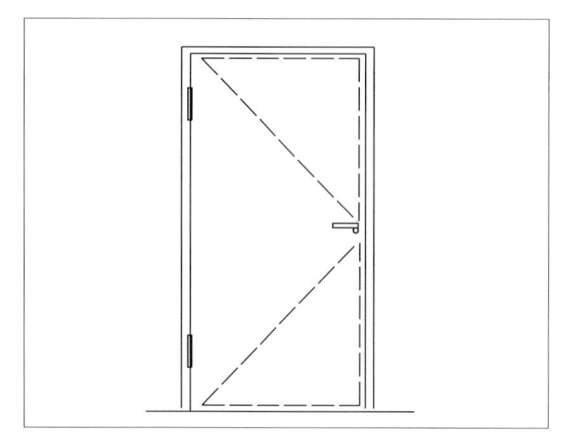

Figure 1-25 A typical door is supported at three points, the minimum needed for stability in the closed position. Large areas on the latch side are poorly supported. If nothing else mattered, a triangular door opening would provide better support for door edges.

Weatherstripping: Most available weatherstripping functions imperfectly when exposed to very low service temperatures and ultraviolet light; flexible plastics become brittle and eventually shatter. Some metal profiles deform permanently under constant use. The positioning of the weatherstripping in the frame relative to the outdoors is crucial. If the strip is placed too far toward the outside its temperature falls below the dew point of the indoor air and hoarfrost begins to accumulate. Placed too close to the wear and tear of foot damage and dirt accumulation, the weatherstripping will deform. The design of the frame, the weatherstripping, and the door must go hand in hand.

TELLTALE

In any building, occupants operate doors more than any other installation. The convenience or inconvenience of doors makes a lasting impression. Doors in bad condition signal a building poorly maintained and badly designed. They are also the most sensitive indicators of differential settlement in a building. By their rectangular (that is, untriangulated) nature door openings have little resistance to lateral strain: frames "rack" and the doors inside them tend to jam shut or jam open. At high latitudes, where differential settlement is commonplace, doorways give early notice of foundation problems.

26 ROOFS

TRADITION

Snow house, summer tent, tipi: all had roofs that looked and performed like walls. Only the immovable houses—the sod house, or the house made of field stone topped by skins stretched over whale ribs—had roofs distinguishable from walls.

PROBLEM

Twenty and thirty years ago many flat roofs built at midlatitudes in North America failed prematurely in spectacular ways. The roofing membrane might develop large cracks or blisters, or it might simply blow away in a high wind. The trouble started when architects and builders began to use traditional roofing materials in new ways without considering the consequences. Even worse, as the flat roof became commonplace it literally dropped out of sight. Unlike a sloped roof, it could not be seen from the ground. So the flat roof dropped out of mind as well.

A perfectly horizontal roof surface exists only in the imagination. In practice, structural decks sag between supports, creating depressions that put perfect drainage out of reach. As structural decks deflect further under the extra load of ponded water, roof drains placed near columns for convenience (that is, where the deck does not deflect as much) remain high and dry. Ponded water, cycled through freeze-thaw many times in the course of a winter, causes expansion and contraction in the built-up roofing (BUR) greater than the membrane's ability to adjust. Cracks open. Insulation placed between the BUR and the heated interior makes the BUR work at lower service temperatures in winter: the membrane becomes more brittle. Poor or non-existent air/vapour barriers below the insulation allow vapour to migrate into the insulation. Just below the BUR membrane the trapped vapour freezes and thaws in unison with the ponded water, destroying the insulation matrix. The presence of vapour induces blisters between the felts of the membrane and accelerates rot of the organic fibres present in the material. Roofing debacles of this kind have cost owners and insurers hundreds of millions of dollars in Canada alone.

At high latitudes in Canada designers and builders imitated the midlatitude example, with the same result occurring in half the time. The fallout from changing the service environment of the roof without changing the materials influences roofing design to this day. To make matters worse, roofing work on new buildings in remote settlements typically occurs in late fall or early winter, when precipitation, high winds, and low temperatures make outside work miserable.

There is a second major class of problems: house designs imported from midlatitudes having sloped roofs (which are effective) with ventilated attic spaces but no air barrier system at the ceiling (which are not). The problem is twofold: blowing snow enters the ventilation openings of the attic and melts in the spring, causing rain showers inside the house; the polyethylene vapour barrier in the ceiling, frequently penetrated by electrical fixtures and access panels,

Roof directs snow and water onto porch. Kimmirut—April

Cold attic with melting hoarfrost. Nanisivik—May

fails to act as an air barrier system because continuous structural support is lacking. Propelled by stack effect and wind pressures, air and water vapour escape upward through the ceiling in great quantities over the winter, causing hoarfrost to accumulate under the roof deck. There the heat of spring sunshine melts the frost, causing a flood to percolate into the living space below.

The first two classes of problems can be blamed on poor design. A third class results from poor roof maintenance. A roof exhibiting minor problems in its first year of life may perform adequately for many years if the maintainer makes immediate repairs. Since roof leaks, followed by roof complaints, are transitory in nature, the maintainer hopes that the effort to rectify the fault can be postponed. Again, casual inspection from the ground reveals nothing about a flat roof. Add to this the maintainer's lack of training in roofing technology and the frequency of problems ceases to surprise. For instance: a BUR originally applied with hot asphalt must be repaired with hot asphalt, even if it is more convenient to apply "cold" solvent-based asphalt. The solvents in cold asphalts attack the chemistry of the "hot" asphalt, rendering it brittle and liable to cracking. The cracks in turn expose the felts in the membrane to moisture that induces rot and, finally, disintegration.

CONDITIONS

Wind: Gales are severest where the land offers little surface resistance to airflow, or where the topography channels airflow from higher altitudes to low-lying areas. Areas below glaciated valleys often sustain high winds. Hurricane force gusts may occur with little warning. In Pangnirtung, steel cables slung over roofs anchor older houses to their foundations. In Iqaluit, high winds remove a roof or two every few years. Suction forces, created by high speed airflow, damage roof overhangs, corners, and ridges in the same way that moving air gives lift to airfoils. Here, gravel on a BUR may be scoured away, exposing the membrane to damaging ultraviolet light; asphalt shingles on a sloped roof may be blown off singly or in groups. Blowing snow—extremely fine-grained snow—frequently penetrates the junction of roof and wall, coming to rest inside the roof assembly, where it may spend the entire winter intact until thawed by spring sunlight. High winds may blow water ponded on a flat roof up and over edge or penetration details whose height above the membrane does not exceed 800 mm. Finally, wind in high latitude settlements blows dust from unpaved streets onto roof surfaces. Accumulated dust retains water, which then refreezes, reducing the flow of water off the roof by constricting roof drains and gutters. Dust also alters the colour and reflectivity of the roof surface.

Radiation: Despite low angle sunlight in summer and sunlight shortage in winter, daylight affects the durability of roofs at high latitudes. Summer days being long, the roof surface absorbs radiation around the clock. Although low angle sunlight has minor impact on horizontal surfaces, adjacent vertical or steeply sloped surfaces catch sunlight at high angles of incidence, and these surfaces reradiate infra-red radiation to the horizontal surfaces. In such an area the surface temperature of the membrane may exceed +50°C. Compare this to the -40°C encountered in winter. Such a variation in temperature causes more expansion and contraction than most membranes can stand. Ultraviolet light from the sky damages the molecular structure of most exposed plastic, rubber, and asphalt roofing materials. The resulting hairline cracks, sometimes accompanied by dust formation as the membrane surface breaks up, signal a material about to lose its impermeability.

Precipitation: Low rates of total precipitation at high latitudes notwithstanding, the accumulation of water on a roof indicates major trouble ahead. Snowdrifts on a building also mean potential roof leaks. Even in areas of the country accustomed to light snow loads the snow swept by the wind from high roofs to lower roofs lying in the lee of the mass of the building may accumulate to depths of several metres. Building codes demand stiffer roof structures to compensate for the extra load. But this does nothing to solve the attendant waterproofing problem. The snow at the bottom of the snowdrift melts from the bottom up in spring, turning to slush. As though in a sponge, the meltwater rises in the snow to heights that may exceed the overlap distance of shingle roofing, sheet metal roofing, or the top of flashing at penetrations, roof/wall junctions, and parapets. The meltwater flowing down the roof through the slush encounters the freezing eaves and refreezes cumulatively. This ice buildup begins to act like a dam, until the meltwater backs up to a point higher than the standard overlap distance of shingles. Any water flowing over the dam at the eaves drips to form icicles. Contrary to expectations steep gable roofs accumulate snowdrifts in the lee of the ridge. The snowdrift adheres to the lee slope until dead weight, structural vibrations, or rising outside air temperatures overcome the bond, releasing the snowdrift in a bruising avalanche.

Air and water vapour: The pressures of stack effect and air turbulence around buildings tend to drive air and water vapour originating inside the building outward through the roof. Water vapour and the higher air temperatures at the underside of the roof promote rot in wood.

Heat transfer: Excessive heat transfer inward from the sun in summer and outward from the building in winter must be moderated by the roof to keep inside temperatures both in the human comfort range and well above the dew point temperature of the inside air.

Surface traffic: Roofs must also survive casual damage. In Alaska and northern Canada the sight of maintainers, thigh-deep in snowdrifts, shovelling snow off a leeward roof to reduce the load is not uncommon. Shovel damage frequently leads to roof failure. Ravens also attack certain types of roofing materials. They seem to favour urethane foam roofing, leaving deep gashes from countless pecks. They also pull silicone sealant from exposed joints.

Drainage: A heavy stream of water draining from the roof will erode the gravel pad foundation supporting one side of the building, accelerating the settlement of the building at that side. Even where the foundation does not include a gravel pad, excess water can contaminate the soil below and around the building, thereby increasing the risk of frost heave.

Vandalism: Where low roofs and high snowdrifts give children easy access, casual play on the tops of buildings may turn to vandalism. Snowmobile traffic on such roofs has predictable results. Even without access, roofs may collect thrown stones. Stones on the roof combine with windblown dust to make a frozen aggregate capable of blocking roof drains.

STRATEGIES

Longevity: Designing a roof to last and building it in mild weather seem obvious steps to take. It is especially relevant in remote communities where roof repairs and roof retrofits will always be expensive. The longer roofing materials stay on the roof the better.

A sloped roof replaces a failed flat one. Kugluktuk—August

Installing MBM roofing. Rankin Inlet—October

Identity: The shape of a roof affects everything from the users' perception of the building to the amount of snow drifting around the building. In a remote community a large roof may be seen for great distances and act as a beacon for homing travellers on the land and for pilots in the air.

Positive drainage: Slope all roof surfaces to falls, either toward the eaves or to roof drains (but not onto adjacent roof surfaces lower down). Given the imperfections of leveling the building during construction and the possibility of differential settlement thereafter, make the minimum slope 1:25 for membrane roofing, and much steeper for segmented roofing. Divert rainwater flow away from the foundation; in other words, design roof drainage and site drainage together. For the safety of passers-by, avoid steep roof slopes in regions where high winds form snowdrifts on leeward exposures, or orient the roof ridge parallel to the prevailing wind to minimize snow accumulations on the roof.

Cathedral ceiling: Abandon the concept of the vented attic space. Adopt the cathedral ceiling so that all components of the roof assembly work together in a compact space with few voids. Where the design of the interior space requires a suspended ceiling, ventilate the space above it to the occupied space below.

Roof deck and ceiling: Select materials that provide rigid support for roofing membranes. Apply preservative to wood members enclosed by a continuous roofing membrane and a vapour barrier or air/vapour barrier. Note that tongue and groove wood decking exposed to the relatively dry interior space of the building shrinks severely across the grain—enough in places for tongues and grooves to disengage—reducing the deck's utility as fire break and structural support.

Air barrier: Provide a continuous, fully supported air barrier, preferably at the ceiling side of the assembly. A barrier located on the ceiling side should join the wall air barrier at the wall/roof junction. A barrier placed on the roofing side should lap but not join the air barrier in the wall: this allows vapour traces trapped inside the roof assembly to escape via molecular diffusion through the unsealed lap joint.

Vapour barrier: Provide a continuous vapour barrier on the ceiling side of the assembly. This barrier should lap the vapour barrier in the wall such that any moisture trapped above the roof vapour barrier drains through the wall assembly, outside the interior wall finish. Allow the vapour barrier to act as a backup membrane in case of a leak in the roofing membrane by sloping it to drain at the exterior wall. A vapour barrier membrane laid on a flat deck will collect water until the membrane is breached.

Vapour escape: Vapour that reaches the inside of the roof assembly by whatever means must be allowed to escape to the outside, laterally to the roof edge if the roofing membrane is continuous and impervious to vapour; laterally and vertically for segmented roofing materials. This is especially important where wood is present. Avoid vapour restrictive compartments between roof joists, for instance; vapour traces should be permitted to diffuse laterally as well as outward from any concentrated source.

Thermal resistance: For small buildings make the thermal resistance in the roof greater than that in the walls to compensate for the warmest air inside being concentrated near the ceiling, but not less than RSI 7. Note that snow insulates. Parapets 300 mm higher than the roof surface may attract a uniform thickness of windblown snow accumulation. Apply insulation in a way or pattern that reduces thermal bridging at structural members.

Weatherproofing: Select materials that shed water efficiently; that endure attack by ultraviolet light; that tolerate a 100 degrees Celsius temperature range in service; that resist wind uplift; that require simple installation and repair methods. The material selected should have a successful track record in service at high latitudes: do not experiment with new materials.

Eaves, gutters, leaders, drains: Where overhanging eaves exist, anticipate ice damming and backflow of meltwater by ensuring the seamlessness of waterproofing along the eaves; design the wall/roof junction so that passive heat loss from the building keeps the eaves relatively warm. Consider eliminating roof overhangs altogether to avoid ice damming, and take other steps to protect windows and exterior wall finishes from deterioration due to periodic wetting. Snowdrifting blocks gutters and leaders installed in regions beyond the tree line, rendering them useless. Within the tree line, despite a lower incidence of snowdrifting, meltwater refreezing on contact with the freezing surfaces of gutters and leaders stops the flow of water to the ground. A half-round section leader performs better than a round pipe in this case; like a gutter, but upright, it guides water flow to the ground and protects the exterior wall surface from wetting, but does not constrict flow at near-freezing temperatures.

Penetrations: Restrict the number of penetrations in roof assemblies to minimize potential for water, air, and vapour leaks. Where penetrations must occur ensure that a vapour-proof collar joins the main vapour barrier securely.

Skylights: Skylights often perform poorly because of leaks and dripping condensation when designed and installed as afterthoughts. Improve the chances of success by ensuring efficient drainage around the opening (steep slopes, crickets) and continuity of the vapour barrier and roof membrane. (A skylight placed at the peak of the roof sidesteps most of these drainage hazards.) Do not place skylights in areas of the roof where snowdrifts will form.

Wall/roof junctions: Since different tradesmen may install the wall and roof barriers on different days, ensure extra care in lapping and bonding of barrier layers at wall/roof junctions. Both designer and builder must clearly understand which layer is doing what, so that the appropriate joint is sealed. For instance, the vapour barrier in the roof should join the vapour barrier in the wall with a watershedding lap and seal. The construction drawings must show this lap clearly.

Accessibility: Provide means of access to all roofs for the safety of maintenance personnel. Where applicable select membrane materials that provide safe footing in all seasons. Some membranes (such as PVC) are so slick that they become a safety hazard in winter, even on a low-slope roof.

Maintenance: All roofs, even new ones, should be inspected in spring and fall for signs of deterioration. Drains and drainage paths must be cleared of debris. In the building operation and maintenance manual handed to the owner at commissioning include a chapter on roof maintenance that indicates steps to take and materials to use in making small repairs. Keep a record of inspections and repairs.

ROOF ASSEMBLY TYPES

There are countless roofing materials available today and many ways (most of them incorrect) to put them together in a roof assembly. Some of the best and some of the worst are identified in the Appendix: modified bitumen membrane (MBM, in *conventional roof assembly*, Figure 1-26), built-up roofing membrane (BUR, in *inverted roof assembly*), sheet metal roofing, asphalt shingles.

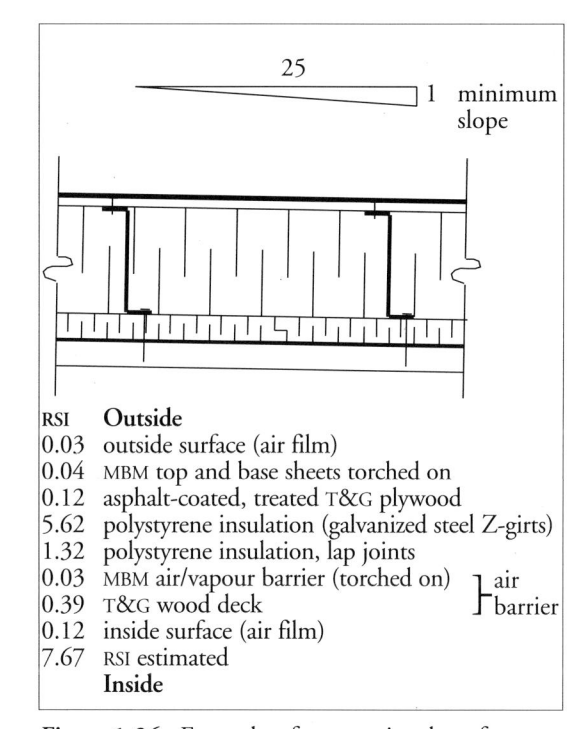

RSI	Outside
0.03	outside surface (air film)
0.04	MBM top and base sheets torched on
0.12	asphalt-coated, treated T&G plywood
5.62	polystyrene insulation (galvanized steel Z-girts)
1.32	polystyrene insulation, lap joints
0.03	MBM air/vapour barrier (torched on)
0.39	T&G wood deck
0.12	inside surface (air film)
7.67	RSI estimated
	Inside

Figure 1-26 Example of conventional roof assembly for high latitudes with modified bitumen membrane (MBM) as roofing material and as air/vapour barrier.

ROOF INSULATION

Glass fibre/mineral wool: Cheap, effective insulation between joists; thermal resistance drops radically when wet. Compact packaging format results in low shipping costs. Loses loft (and thermal resistance) over time when installed in roof assembly.

Wood fibre board: Comparatively low initial thermal resistance; deteriorates completely when wet:

moisture damage spreads laterally for long distances from a single membrane leak.

Low-density polystyrene boards: Absorbs some moisture at and near surface when wetted. Available with factory-made slopes. Disintegrates on contact with propane torch or hot asphalt.

High-density polystyrene boards: Good thermal resistance and moisture resistance; also suitable for inverted roofing when covered by protection sheet and clean gravel ballast.

High-density phenolic foam boards: Stated thermal resistance may deteriorate considerably over time as formation gasses escape from the plastic matrix. (Check independent test results.) May promote corrosion of adjacent metal in certain assemblies. Has tendency to warp when heated.

MATERIALS

TRADITION

A rock rolling down a slope releases potential energy. A well-built cairn conserves this energy. Suppose for a moment that this energy could be transformed one night into the red glow of hot coals. The barren ground shorelines would light up like ruby necklaces, each hot spot representing an act of human diligence.

People raised those piles of stone. Their cairns, caches, traps, and foundations are quite unlike any rock pile left by receding glaciers or spring floods or freeze-thaw. They stand out against the light. Each rock is small enough to be handled by a human working alone but large enough to resist overturning by wind for centuries at a time.

INDIGENOUS MATERIALS

Field stone: The first mission fathers to work in arctic communities also appreciated the permanence of rocks. They had the walls of their churches built of stone gathered from the beaches by devoted congregations. The rock was neither cut nor dressed; the masonry was rough. No heater was big enough, no mission fuel-rich enough, to fill those massive walls with heat. Worshippers wore winter clothing to prevent the stonework, and the wind coursing through it, from snatching body heat during services. Today, congregations rarely erect public buildings on their own. Masons remain scarce; the exterior masonry season is very short; suitable stone for quarrying may not exist locally or may lie beyond the reach of the road. For a building suspended above permafrost, stone walls may weigh too heavily and expensively in the foundation load calculation. But stone may have other uses: a local exposure of bedrock may weather in a useful way, producing flagstones for a floor or pathway. All that is required is the will to gather and transport them to the building site. To the community the symbolic importance of including local materials in the construction of a new building may outweigh the cost.

Gravel and sand: Gravel and sand materials, too heavy to carry, had no place in the architecture of hunter/gatherers. But by being well-drained and exposed to sunlight, gravel beaches offered a permafrost table low enough to make shallow excavation in summer practical. Houses could be set half-way into the ground to permit cold-trap entrances and reduce the amount of wall and roof to be made up in skin and bone. In northern Canada natural gravel and sand are now used extensively for roadbeds, airstrips, and pads on which to set buildings, and, secondarily, as aggregate in concrete. Demand frequently outstrips supply. Since near sources are exploited before distant ones, quarrying disfigures the landscape adjacent to the settlement. The price of the material from a sole source in private hands may be exorbitant. The source is often contaminated by sea salt, prehistoric or modern, limiting its potential as aggregate in concrete. Given the climate, the source is likely contaminated also by frost.

Crushed gravel, and the machinery to make it, are rarer still. Rock available for crushing may fracture in ways that produce low quality aggregate for specific uses, say for high strength concrete. From time to time the construction or improvement of an airport runway in northern Canada brings rock-crushing equipment to a remote community to produce granular material. The equipment is used to stockpile gravel for government use and then is shipped away to a new location. In West Greenland the long shipping season permits barging stockpiled gravel from sources near towns to outports along the coast.

Sod and peat: Scarcer still than gravel and sand, sod and peat were used to build walls and sometimes roofs of houses wherever the material could be found in quantity. A few sod houses still stand. Unfavourable comparison with modern materials has eliminated demand for sod houses, and the skills needed to build them have disappeared.

Skin and bone: The availability and portability of animal pelts was crucial to the architecture of hunter/gatherers. The useful life of pelts rarely exceeded three years, however, so the demand for replacements never abated. Whale bone and caribou antler for roof structures lasted longer, but were not portable. Microbes and insects eroded these materials constantly. Many animal sources of the material are now rare, protected by legislation against human predation.

Logs: Driftwood propelled northward to the sea by the Liard, the Slave, and the Mackenzie Rivers and others have provided windfall material for tools, shelters, and firewood for millennia. In recent centuries the windfall was spiced by flotsam from wrecked whaling vessels. Uncertain supply limited methodical use. Toward the middle of this century

Thule house ruins. Skraeling Island, Ellesmere—July

the move from aboriginal housing to the drafty shacks invented by outsiders caused the supply of driftwood to disappear into the bellies of wood-burning stoves. Within the tree line, fresh cut logs rafted on the river and seasoned make useful building material. In subarctic Canada a few communities remain whose buildings are almost exclusively made of softwood logs, logs for walls and logs for roof structure. Log work there may be inseparable from community culture.

Softwood logs, usually spruce, in northern Canada, have only a quarter of the thermal resistance of plastic foam or dense glass fibre insulation. This disqualifies the use of logs as the primary wall insulator of modern houses in most jurisdictions. The use of log walls in larger buildings, whose surface-to-volume ratio is lower than that of a detached house, is feasible, but there are other inherent problems. Logs shrink along the grain and shrink proportionally even more across the grain under the drying effects of sunlight and central heating. Without an air barrier system made of other materials a log wall behaves like a sieve. Significant numbers of the logs harvested in northern Canada have spiral grain. Individual trees, responding to site-specific wind effects and asymmetric branch development, grow upward in a slow spiral. The grain is twisted. A crack in a log with spiral grain develops spirally. The crack may begin outside the building and migrate with the grain to the inside. The result is a major air leak.

Snow: Air trapped between crystals of snow accumulating on the ground makes snow an excellent insulator; dense snow makes an excellent building block. Snow's transient nature and low service temperature limit its everyday usefulness to emergency shelters for overland travel and auxiliary walls intended to deflect wind from the leaky porches and crawl spaces of poorly designed wood houses.

MODERN MATERIALS

Manufacturers of building materials aim their wares at large and expanding markets; they locate their factories between the source of raw materials and the places inhabited by the most customers. Since the biggest markets exist at midlatitudes, manufacturers develop products to suit midlatitude buyers and environments. High latitudes, with small, remote centres and climatic extremes, are perpetually excluded from this commercial stratagem. For lack of building products perfectly suited to the high latitude environment the designer must choose between "the next best thing," a material that fails early but is easily replaced, and a radically different construction detail designed to neutralize the negative qualities of inferior products.

Building materials purchased at midlatitudes for service at high latitudes must run the gauntlet of multiple-handling, several means of transport, exposure to air, sun, sea, rain, dirt, and sometimes frost over thousands of kilometres. Once at the site they are stored in the open, perhaps over a winter, since remote communities possess little warehouse space. The stress of transport on the material may be greater than the stress of life in service. Unfortunately, some materials recover from the stress of transport only after reaching their niche in the building assembly: as they adjust to the new environment they distort the building irreversibly.

PERFORMANCE

Performance selection criteria for building materials spring from understanding function, simplicity, durability, appearance, and cost. The "best" material is rarely the best performer; material selected must perform compatibly with adjacent materials in the assembly. The criteria most often in focus are listed on the next page.

DEGRADATION

Building material performance tends to peak inside the manufacturer's warehouse rather than inside the new building. While manufacturers extol peak per-

MATERIALS 143

Maximize:	**Minimize:**
service life	number of different materials
durability	number of different trades required for installation
dimensional stability	initial cost
utility-to-weight ratio	operating and maintenance cost
compatibility with adjacent materials	weight
speed of assembly	shipping volume
ease of field-modification	moisture content
ease of repair and replacement	fire fuel contribution
quality of appearance	flame spread
specific heat (heat storage capacity divided by the mass)	toxicity (for instance, pentachlorophenyls, inorganic arsenic)
number of sources (to improve price competition)	off-gassing
certification (by, for example, the Canadian Standards Association)	last-minute substitutions

formance, the building designer has to consider a material's lifetime performance. What good is an insulation that has an initial thermal resistance 30 percent higher than its competitor if its resistance falls off 30 percent in two years because of loss of foaming agent gas by diffusion through the insulation matrix? The owner pays a premium for that initial 30 percent extra. How does the departing gas otherwise affect the building, its occupants, and the outside environment?

Once off the manufacturer's assembly line the material is subjected to new environments. The air pressure is one atmosphere, for instance, instead of three atmospheres: the material expands or, losing fluid and gaseous constituents, shrinks. The air temperature is -40°C instead of room temperature: the material contracts. The relative humidity outside is 75 percent instead of 30 percent: the material expands or corrosion accelerates. Water in or around the material freezes and expands, thaws and con-

tracts. Exposure to ultraviolet light is quintupled: the molecular structure of the material begins to change. Degradation is irreversible. Just as rubber bands lose elasticity with age, useless properties replace useful properties in building materials.

PREFABRICATION

Prefabrication of building components surfaces from time to time as the ultimate solution to the world's intractable shelter shortage. The impetus for the idea often dies prematurely or, where totalitarian authority holds sway, prefabricated buildings conceived in haste survive to become slums in less than a generation. Like junk yards, defunct prefabrication projects litter the developing regions of the earth. The world shelter shortage is not a technical problem: its roots are political, social, and economic. At high latitudes prefabrication has succeeded only where money is no object, or where a community is a one-company resource town: in other words, where every other

social and technical consideration gives way to achieving a high level of productivity in the shortest possible time.

CONCRETE

Most remote communities in northern Canada cannot afford to keep batch plants and cranes for mixing and placing concrete. Cement and reinforcing steel must be imported and local gravel stocks are often too short, poorly graded, or contaminated. Concrete floor-finishing equipment must be imported for each project. Concrete is used sparingly for these reasons; typical uses are cast-in-place floors for mechanical rooms and garages in public buildings, and foundation walls in discontinuous permafrost areas having well drained soils. Accidents such as irregular curing, or allowing the bearing soil to freeze, still occur. The casting and curing of concrete uses large amounts of water, much of which evaporates: it is essential that permanent enclosures be protected against water vapour exfiltration by having envelope air and vapour barriers installed prior to pouring concrete or by pouring before the structure is enclosed. Allowance must be made for the movement of structural concrete due to shrinkage and creep.

MASONRY

The high incidence of breakage and high shipping cost discourage the use of clay, glass, and concrete masonry units at remote high latitude sites. The expense is compounded by importing masons, specialists who cannot perform any other task. Concrete block walls shrink with age and, contrary to appearances, are useless as air barriers. Clay brick masonry units tend to expand with age as they take on moisture from the air. Differing dimensional changes

Prestained, multi-coloured wood siding. Resolute—September

between masonry and supporting structure often results in masonry wall failure. Efflorescence—the appearance of salt precipitates on the exposed face of new masonry—while understood in theory, still bedevils practice. Masonry walls unprotected by projecting eaves may degrade quickly through wetting and freeze-thaw cycles.

METALS

The use of steel structural members for large buildings, or small buildings with large spans, is common in the remote communities of northern Canada. Most settlements are located within two kilometres of salt or fresh water surfaces where threshold humidity (75 percent rh) conditions for corrosion of exposed steel persist. Given its high coefficient of expansion, steel must either be placed in a thermally stable environment (fully inside the heated building, for instance, not halfway through the envelope) or detailed in such a way that adjacent materials do not interfere with expansion and contraction movement. Steel's high thermal conductivity makes it a potential thermal bridge.

High thermal conductivity limits the usefulness of pressed steel and extruded aluminum for exterior window and door frames: too much heat is lost to the outside, causing condensation on the inside. Thermal breaks placed by manufacturers inside metal frame sections to counter heat loss can be rendered useless by careless placement of the frame within the opening: heat may flow around the thermal break through other materials in contact with the frame.

Designers often select steel sheet to clad and roof utility buildings at high latitudes. It is cheap to install and maintain. Since steel sheet expands and contracts significantly in response to abrupt temperature changes, allowance must be made for cyclical movement to prevent buckling or "oil canning," from which there is no recovery. Motor vehicles frequently inflict impact damage on steel sheet cladding. Snow accumulation around buildings raises vehicles above road level, permitting contact with walls at shoulder height. Although the technical damage may be slight (repair means costly sheet replacement), it certainly ages the building in the eye of the beholder. By throwing shade into the slightest of depressions, low-angled sunlight picks up every dent in the cladding.

WOOD

Wood by any measure is a remarkable building material. It is a natural insulator—a poor conductor of heat. Wood has a high utility-to-weight ratio; it is easy to work; it is familiar, even in remote communities having no trees taller than five centimetres; it is replaceable; it is handsome in colour and texture; it may offer a pleasant smell, sometimes long after the building has been opened. Unfortunately, the price of wood steepens as midlatitude stocks of commercial-grade forest decline.

Wood is porous and dimensionally reactive to drops in moisture content and ambient relative humidity: it shrinks, cracks, and splinters. At remote sites wood often arrives in a new building thoroughly soaked, and the subsequent shrinkage deforms other attached materials. At ambient relative humidities exceeding 30 percent wood expands. Wetted, it expands even more, damaging adjacent materials. In the presence of moisture, oxygen, and fungus, wood decays at temperatures above +5°C that normally occur in a heated building. At still higher temperatures the material burns freely.

Wood, factory-glued and laminated to form custom-designed beams and columns, performs well provided that members remain in a thermally stable, moisture-free environment.

Exterior wood cladding, left to itself, weathers moderately well but becomes brittle under the constant wetting/drying and low absolute humidity conditions prevailing outside. Its life can be usefully extended by applying a sealer or other vapour-permeable coating. Applying sealer at or near the lumber source prior to shipment to the site sidesteps seasons too cold for field-application of coatings, and one round of scaffolding.

PLASTICS

For fifty years chemists have showered the construction industry with a bewildering array of special-purpose plastics. Since the industry cannot fully integrate new materials without first having them tested in laboratories and in field conditions, the evolution of plastics in building has been chaotic. Survival of the fittest material is complicated by spotty market research, competition with cheaper established materials, company diversification and marketing strategy, and formidable start-up costs for assembly lines. The life of a plastic product on the market may be ten times shorter than its service life in a building, thereby complicating future repair and replacement.

The failure of certain roofing membranes, insulations, and window frames and the abrupt withdrawal or restriction of others (urea formaldehyde and polyurethane insulations) in recent years now seems to have been inevitable. The impact of weathering, particularly the ultraviolet component of sunlight, alters the molecular structure of the plastic: useless properties replace useful ones. (The loss of special properties may be rapid but the residue may persist for generations: witness the wind-blown proliferation of disposable diapers and foam insulation off-cuts on the tundra.)

Open-celled foam insulations absorb water and water vapour easily; closed-cell foam insulations absorb much less. Some foam insulations are more susceptible to flame spread and burning than others; all are dimensionally unstable—they expand and contract significantly with large changes in ambient temperature. Clear acrylic skylights and window panes require special edge details to accommodate this movement. Vinyl floor tiles shrink with age.

GYPSUM BOARD

In northern Canada gypsum plasterboard—drywall—predominates interior wall and ceiling finishes because it is cheap, simple to apply, easy to paint and repaint, moderately resistant to mechanical damage, and highly fire resistant compared with alternative materials. The major drawback of drywall is its inability to resist or accommodate structural movements without special measures. Buildings erected on frost-susceptible soil or built with structural materials (wood) that change dimension when wetted or dried are bound to move differentially. Without allowances, drywall covered with paper tape and plaster at panel joints and intersecting planes will crack. A gypsum board ceiling coated with white paint looks striking until shrinkage in the roof structure riddles the visible surface with a jagged network of cracks. Allowance for building movement can be made by installing metal control joints (offered by the gypsum board manufacturer) wherever appropriate but at intervals less than ten metres long. Where drywall meets an intersecting plane or another material, an open joint (defined again by metal or plastic strips sold for the purpose) that allows the gypsum panel to slip when stressed is simple to detail and install. Another alternative is to install prefinished gypsum boards whose panel joints are either left open or covered with a metal batten.

COATINGS

Common coatings consist of a solvent or water base that carries a binder to fix the dry film to a surface, plus a pigment (insoluble) or a dye (soluble) that provides colour and hiding power. Without pigment or dye the coating is transparent. Solvent-based coatings (oil paint) release noxious gases as they dry; water-based coatings release mostly water. Solvent-based coatings in liquid form are frost-resistant; water-based coatings in the can are not. This distinction counts where transhipment delays may leave materials exposed to freezing weather. Two coats of a solvent-based coating can act as a vapour barrier, although there is no guarantee that the substrate itself will not crack, permitting passage of vapour and air. (A new dry film is extensible up to about five percent of its length.) Water-based coatings are vapour permeable. Sealers, with little pigment to provide a thick continuous dry film, are also vapour permeable. The coating selected for exterior surfaces must not block the passage of vapour to the outdoors.

Typically the interior painting stage falls in late winter, when the temperature differential across perimeter walls is steep. Near the inside face of the perimeter wall the differential causes an electrostatic potential attractive to airborne dust. The potential is greatest where a wood or metal stud creates a thermal bridge through the wall. The result is "*pattern staining.*" A pattern of dust particles accretes on the coating in the image of the underlying studs. Dust accretion over thermal bridges continues long after the coating has dried. The worst effects of pattern staining can be avoided by applying the interior coating in warm weather, when the temperature differential through the perimeter wall is low. Given low maintenance standards at high latitudes, the durability and washability of interior coatings must figure in material selection: a coating must last longer between recoatings.

COLOUR AND TEXTURE

A readiness to accept colour schemes deemed least controversial to the most people marks the selection of colour in northern Canadian buildings. Colour in shelter, other than the colours of indigenous materials used in construction, has little tradition in aboriginal culture in northern North America. Instead, the artistic use of striking colour, pattern, and texture was reserved for clothing, particularly those pieces of clothing worn on special occasions. In good times individuals put on their best colours, their own choice. Colouring people was more appropriate than colouring shelter. Euro-Canadian culture, while less homogeneous than those aboriginal societies, seems to favour the use of colour in buildings as well as in clothing.

Many building designers support their preference for striking colours in building by emphasizing the need to counter-balance the cool, drab colours of the arctic landscape. They suggest that bright, primary colours boost morale during the long, cold winters. Most northerners in Canada do not consider the landscape they live in to be dull, however, or their morale to be in need of boosting. Given the glut of colour picture books about polar regions at midlatitude book stores the "drabness" theory seems feeble.

Too often building designers hire the services of an interior design specialist late in the design phase of the work. At that stage the interior designer can do little more than "dress" the spaces that have emerged. If involved from the start the interior designer can bring into sharp focus matters of space planning, finishing, and detail that otherwise might remain partially resolved at best.

SERVICES

TRADITION
Build compact shelters to keep demand for "services" low. Open the hole at the top to let fresh air in at the bottom; close it for warmth. Gather enough deadwood to burn tomorrow; let the residue return to the land. Drink water upstream; pollute downstream. Waste little; leave the rest to the dogs. Drop rejects anywhere outside; move camp to a fresh campsite. Shut out the spring sun with goggles; let pupils dilate to gather the dim light of polar night.

IN THE FIRST PLACE
People who spend most of their lives indoors want to control their surroundings: the air temperature, humidity, noise, air and water flow inside their own homes and places of work. Too often, at high latitudes, they cannot. The heaters overheat, the windows freeze, the doors stick, the humidity rises sharply or dips so low that skin turns to paper, combustion air vents let in fine snow. The maintainer's refrain: "nothing can be done about it." Minor maladies can always be blamed on other causes, and relief in the form of summer and open windows is never more than six months away. The incentive to examine and correct inside environmental problems may be too slight.

Why do building occupants at high latitudes react passively, even resentfully, to chronic failure in building systems? There are several reasons. First, services designers and occupants rarely meet face to face before, during, or after a new building is opened. Designers may never learn how poorly their design performs in everyday use. They are not directly accountable to the occupants. Second, engineers are specially trained and users are not; the two think and speak differently and fail to communicate through the architect. Third, since users are "not to be trusted," engineers increasingly design controls for services to bypass the occupants, to perform unassisted by human hand—to treat users as incompetents. This syndrome has two obvious faults: no human can design a foolproof system, and treatment of users as incompetents perpetuates incompetent behaviour. Fourth, modern controls and any defects in them are increasingly sophisticated, to the point that the original supplier alone has knowledge sufficient to repair an ailing system. The supplier holds occupants hostage. Fifth, modern services interact with building geometry in ways so complex that precise solutions to problems may be beyond the reach of ordinary specialists.

Occupants deserve to be masters of their own environment. To keep the simplicity ethic intact the architect and the engineers must agree on a level of mechanical and electrical services appropriate to the new building.

HEAT GAIN, HEAT LOSS
At high latitudes an unventilated room with picture windows facing the equator marks the work of an indifferent designer. The sun's return in spring turns the room into an airless greenhouse in which both

Floor over an open crawl space, with services roughed-in. Rankin Inlet—October

Kitchen from another culture does not fit local need. Gjoa Haven—February

plants and humans wilt. The thermostat, sensing desert heat, idles the heating system while shaded parts of the house cool below comfort level. On a cloudy day, heat flows out of the room through the expanse of glass, cooling the space and depressing the thermostat. The thermostat demands heat from the boiler and the back rooms now overheat.

At high latitudes exterior walls catch the brunt of solar heat from a sun that hugs the horizon much of the summer day. Solar heat on the outer surface of a wall slows the outward flow of heat through the building envelope, just as a high tide stems the flow of fresh water at the mouth of a river. At a window, transparent and unresisting, the heat flow reverses itself temporarily: more heat flows in than flows out. In the absence of sunlight on a cold, cloudy day or in winter twilight, heat flows outward through windows.

In principle, the volume of air contained in a building can be heated to room temperature very quickly. In practice the building envelope and everything it encloses—furniture, partitions, floors, and ceilings—must also be heated for several hours before the building feels warm to the occupant. Until the furnace or boiler has produced heat sufficient to fill all those materials to their heat storage capacity at room temperature, warmth radiating from a human body will be absorbed, not reflected, by the surroundings. Finally, the materials reradiate about as much heat as they absorb, and the one-way trip of radiant body heat ceases. The occupant feels comfortable quite suddenly. Building enclosures and contents having low heat storage capacity—air, wood, plastic foam—warm up quickly, and lose heat just as quickly when the furnace is turned off. Materials with high heat storage capacity—water, brick, stone, gypsum board—take longer to warm up, and longer to cool. High heat storage capacity, the heart of "passive solar systems," rounds off the thermal peaks and dips of on-again, off-again heating cycles. In a space with low heat storage capacity the furnace or boiler must cycle frequently to keep air temperature in the comfort range.

Water in the human body, water in storage tanks, and water in the vapour phase all contain heat, much more than an equivalent volume of dry air at normal pressure could contain. Every time we exhale we lose some body heat in the water vapour expelled. Every time an outer door is opened in winter a packet of heat leaves with the vapour in the warm air lost through the opening.

STACK EFFECT

Heat flow and air circulation in an enclosed space do not depend on human intervention. Physical principles do the work unassisted, but the result may be uneven, falling only accidentally inside the human comfort zone. In the example of the picture windows exposed to sunlight, air at warm inside surfaces rises while air at cooler surfaces descends. At sunset, the warm air accumulates against the ceiling and the cool air pools on the floor. Occupants, with their heads in the tropics and their feet in Antarctica, develop mysterious headaches. The air has stratified, with warm layers on top, cold layers at the bottom.

The buoyancy of warm air in an ocean of colder air also drives "stack effect" (stack, smokestack, and chimney are synonymous). A person standing in a hot shower can sense heated air leaving the stall at ceiling level while replacement air surges past the bottom of the shower curtain. Stack effect is the engine of natural ventilation in buildings. It is more pronounced in tall buildings than in short ones, and in heated buildings surrounded by cold air than in shaded tropical buildings (Figure 1-28). The invention of the chimney harnesses this principle to create a draft, which expels combustion gases safely above the heads of building occupants and draws fresh air through the base of the fire to improve combustion.

At high latitudes during the cold months the buoyancy of heated air—the stack effect—contributes to the leakage of warm air through the enve-

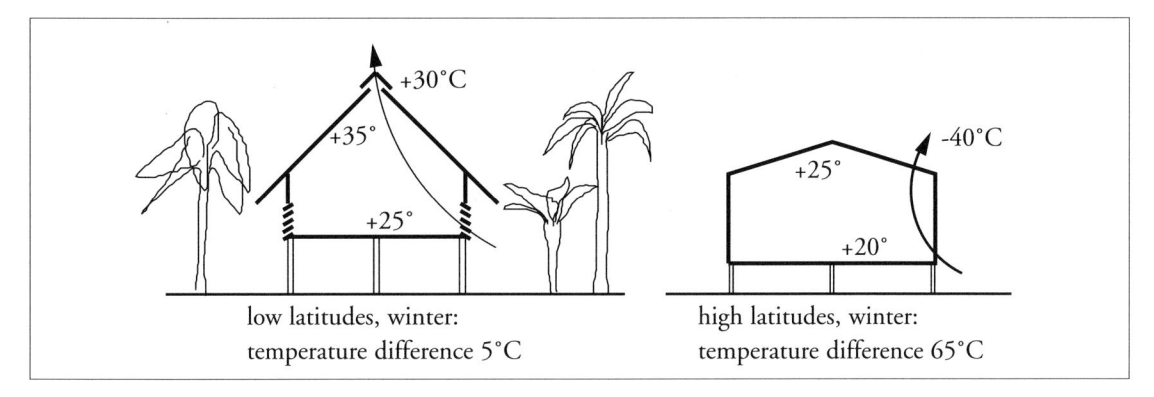

Figure 1-28 Stack effect, driven by the difference between outside and inside air temperatures, is strongest at high latitudes in winter, where the cooling effect inside caused by the exchange of air is least desirable.

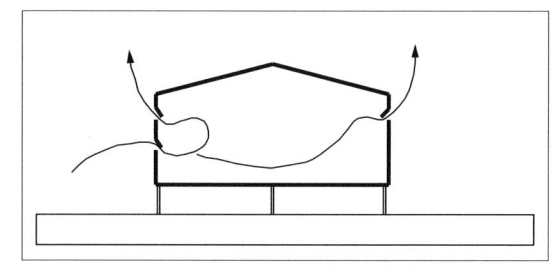

Figure 2-28 Controlled natural ventilation induced by stack effect and strategic placement of operable portholes at high and low points in the perimeter walls.

lope of the upper part of a heated building and, to compensate for the lost volume, infiltration of cold air through the lower part. The greater the amount of warm air lost, the greater the quantity of replacement air to be heated.

VENTILATION

Some high arctic residents with vivid memories of life in snow houses continue to be baffled by the outsider's resistance to placing a small ventilation hole through the roofs of modern buildings. They realize that most new houses contain air too stale or too humid for human comfort. Some tenants have surreptitiously installed ventilation holes in their homes. Complaints about interior air quality in those houses have since ceased. This is natural ventilation controlled by the occupant.

Natural ventilation controlled by the occupant is cheap and effective. In northern Canada, several new buildings put the principle to work by means of controllable vents at roof level or of operable portholes at high and low points in perimeter walls (Figure 2-28). Natural ventilation at high latitudes is not a pipe dream. In large or complex buildings mechanical assistance to natural ventilation (required by code) is laudable provided it does not pervert the aim of creating an enjoyable environment easily controlled by its occupants.

Wind effect around buildings also drives natural ventilation. Wind forced to go around an object creates positive air pressures at upwind walls, and negative air pressures at downwind walls. The result is air infiltration through cracks and openings in the envelope at upwind positions and air exfiltration at downwind positions. Often stronger than stack effect, the wind may temporarily reverse the normal pattern of airflow through vents and portholes.

HEATING

The volume of air contained in a modern high latitude dwelling is ten times greater than that of a tipi, twenty times greater than that of a snow house. Not only must the larger volume be heated, it must be heated to a room temperature 10 to 20 degrees Celsius higher than the room temperature considered normal in northern Canada sixty years ago. The energy to do this extra heating originates in the Gulfs of Oman, Campeche, and Venezuela. Seal oil and deadwood for heating have become cultural relics.

At high latitudes the use of diesel-generated electricity for heating is very inefficient. By the time fuel oil is converted to electricity for electric heating, more than sixty percent of the oil's heat potential is lost. Instead, fuel oil is burned directly either in hot air furnaces or in hot water boilers, reducing the amount of heat potential lost to about thirty percent. That thirty percent goes up the chimney as hot gases and imperfectly burned fuel.

For houses and small buildings the heating system medium can be forced air or circulated water. Forced air means electric fans and fan noise, ductwork too ungainly to be left exposed, and dust-laden air circulated around the space many times a day through an air filter that may not be replaced often enough. The oil-fired furnace is quite simple to maintain. The hot-water heating system is moderately more efficient,

Service entries and exits severely stressed. Taloyoak—February

distributes heat quietly and more evenly around the perimeter, and does not push airborne dust from room to room. The heating medium—a mixture of water and antifreeze—has a heat storage capacity many times greater than that of air, so much less medium has to be circulated to distribute the same amount of heat. The workings of boiler, circulation pumps, and antifreeze, while less familiar to some maintainers in remote communities, offer the best performance in the long run. The steel covers of baseboard radiators resist impact damage poorly, however, and the slight risk of pipe burst and flood due to freezing is unavoidable. Room-temperature water circulated from the boiler through continuous loops of plastic pipe located in shallow floor spaces offers a reliable antidote to cold floors built above unheated crawl spaces.

In some communities in northern Canada the local diesel-electric generating plants have been jacketed so that the waste engine heat previously vented to the outdoors is now recovered and circulated to heat any large building located a short distance away. Many experiments with air-to-air heat exchangers have taken place in northern Canada, the aim being to transfer heat from outgoing stale air to incoming fresh air, with poor results. Frozen condensate infests the equipment when outside air temperatures dip below -20°C. Automatic vents that feed outside air indirectly to boiler combustion chambers often freeze in very cold or blizzard conditions.

WATER

At high latitudes so much depends on water remaining in liquid form—domestic hot and cold supply, hot water radiators, fire protection, sewage disposal—that prevention of ice formation in pipelines and storage tanks takes conscious effort. Water pumped from an ice-covered lake must be heated a few degrees Celsius above freezing before being sent down the pipe toward the settlement. At the settlement it must be reheated again before being distributed or recirculated.

Water pumped from a shallow lake in winter tastes bad. Such water stored in a plastic tank located in a hot mechanical room becomes lukewarm and tastes even worse. There is a risk of biological growth in water at that temperature. Many northern Canadians, preferring to drink lake-ice water, keep a second, smaller tank full of melting ice near the kitchen. The large storage tanks are filled from a tanker truck at uncertain intervals. The driver extends a hose to an infill pipe placed in the wall of the building, engages the tanker's water pump, and then waits until water overflowing from the tank through an adjacent vent pipe signals "tank full." Any spilled water that freezes inside the overflow pipe blocks the release of air expelled from the tank by the volume of new water, causing the tank to explode. Any spillage that runs back along the outside of either pipe seeps down inside the wall, and freezes in the insulation. The rest of the spilled water drops down to the ground outside where, in winter, it forms a stalagmite.

SEWAGE

Where sewage is not collected by a pipe network or stored in a tank, grey water from baths, basins, and sinks drains through an exterior wall to form another stalagmite, which melts odorously in the spring. Bagged toilet waste is collected at intervals in a pickup truck and dumped in a sewage lagoon outside the community. Insulated sewage tanks buried all or partly in the ground may be shifted detrimentally by frost heave; the gradual flow of heat from the tank to the adjacent ground may also degrade the permafrost that supports the foundation of the building. Uninsulated tanks placed inside the building generally work well with the following exceptions: if pump-outs are too irregular for any reason, a blizzard immobilizing the sewage pump-out truck, for instance, the tank may fill to capacity, thereby interrupting service. The plumbing vent from the sewage tank up to roof level may become congested with condensation ice at the top, causing a vacuum in the tank at time of pump-out that might cause it to implode, rupturing pipe connections. Vents cut nearly flush with the roof plane, and so kept relatively warm by heat loss from a cathedral ceiling, perform better than vents with extensions fully exposed to the outside air above the roof.

SPRINKLERS

Automatic piped water fire protection systems—sprinklers—are expensive to install but very effective in preventing fire. The two or three percent failure rate is largely attributable to inadequate operation and maintenance, including the careless closure of valves that should remain open. The installation cost in buildings larger than a house can often be partially offset by reduction in the amount of fire-resistive construction and compartmentalization required by code in unsprinklered buildings. (Since the water inside sprinkler pipes is stagnant it is especially susceptible to freezing if located too close to unanticipated air leaks in the building envelope: keep sprinkler pipes and building envelope apart.) In the absence of street mains, space must be found inside the building to house a large tank of water dedicated to fire protection. Usually this is combined with domestic water storage, so that domestic water use prevents stagnation in stored water. To keep the height of crawl spaces low, under-floor tanks made of large-diameter plastic pipes can be substituted for conventional water storage tanks.

FUELS

The movement of heating oil and propane in small diameter pipes between the outside fuel tank and the heated building becomes sluggish at temperatures around -40°C. Remote communities stock only heating oil, in a tank farm located at the village periphery that includes gasoline and aviation fuel reserves. Fuels are delivered to the tank farm once a year by barge. Heating fuel is distributed by truck and loaded into small tanks located against a wall outside the building, usually at a height to prevent burial under snow, or degradation of the permafrost.

UTILIDORS

In northern Canada the 1950s fascination with *utilidors*, above-ground insulated pipe service networks, has declined with experience. Water in pipes is highly susceptible to freezing in winter, particularly where exposed to cold air drafts. The task of protecting pipes enclosed above ground against cold air infiltration for long periods has proven formidable. Utilidors criss-cross the landscape, blocking foot and vehicular traffic while degrading the view. As the network is extended by different authorities to different standards, debate as to who pays for the escalating cost of maintenance becomes rancorous and time-consuming. Dilapidated utilidor networks outnumber well-maintained ones.

In winter, buried pipe lines are exposed to ground temperatures twenty to thirty degrees Celsius higher than the air temperature at the surface. There are no wind chill effects underground. The advent of pre-bundled and pre-insulated pipe systems has simplified underground operation and maintenance. Less energy is required to circulate and reheat the water that keeps the pipe mains a few degrees above freezing the entire year.

ELECTRICITY

Power is distributed from the diesel generating plant to consumers via overhead cables raised on eight-metre poles set partially into the ground. Where burial of the butt end of the pole is impractical the pole is set into a circular steel caisson resting on the ground and ballasted with rocks. Compared to west Greenland communities, where secondary distribution is laid below ground or at the surface, the apparent disorder of many northern Canadian communities is accentuated by the tangle of overhead wiring strung from poles leaning at odd angles. (Streetlighting, considered vital to safety, blinds residents to the night sky, and so must

A too-sensitive smoke alarm is "socked" by tenant. Rankin Inlet—March

*Refuse basket with disposable bag and cover.
Nuuk—August*

be added to the list of improvements that separate people from the land around them.)

Electrical penetrations can become major air leaks. Installation of wiring and fixtures should be concentrated away from perimeter walls and cathedral ceilings. Alternatively, wiring and fixtures can be installed in a 38 mm uninsulated space formed by furring on the inboard side of the vapour barrier, assuming a secure air barrier system elsewhere in the perimeter wall assembly. Running electrical wire in metal sheaths (wire mould) on the interior surface of perimeter walls risks corrosion of the sheath and rust-staining of the wall wherever windows produce dripping condensation.

LIGHTING

Over the years the attitude of experts has varied widely over what constitutes proper light levels for various human activities (more light is good; less light is good; no, less light better directed is good) and what constitutes the best method for calculating light levels in new buildings. In northern Canada, where the nearest lamp producer is at least a thousand kilometres away, the debate expresses itself in apparently random selection of lamp types and lighting levels. On a sloping ceiling in a school classroom fluorescent lights are aligned in a regular pattern, ignoring the matter of varying distance between lamp and the surface to be lit: the lighting level on desk tops at the low end of the room may be twice that at the high end. A five-year-old ceiling lighting scheme becomes a patchwork of fluorescent tubes with different colours, because the wrong type of replacement tubes were ordered, or the ones originally specified are no longer available.

There is much room for useful innovation. The key is to visualize the activities expected to take place in a given space and to design lighting appropriate to the activity. The factors involved include room surface reflectivity and colour, lamp efficiency (fluorescent lamps are about four times as efficient as incandescents), light levels and distribution, colour rendering, glare, reflections on video monitors, noise level, maintenance accessibility, maintenance reliability, and availability of replacement parts. The fewer the number of different fixtures the better.

GARBAGE DISPOSAL

In Greenland the commune provides the householder with a paper bag that can be inserted into a wire basket with a cover attached, outdoors on the back porch. The basket opens wide like a clam shell for ease of removing loaded bags. Birds and quadrupedal scavengers cannot penetrate such baskets. In northern Canada domestic refuse is often deposited in an open steel drum by the roadside. Scavengers have little difficulty in upsetting the barrel and scattering the contents to the winds—another service problem that has never been properly resolved.

The garbage is collected in a dump truck and sent to the settlement landfill, where it is partially burned and then bulldozed flat. There is no soil with which to bury the deposits, so scavengers and wind have another chance to go over the material. Since present-day northern Canada, like Canada itself, throws out "rich" garbage, savvy residents know that reusable pieces of material can often be found at the settlement dump. So the dump doubles informally as a community "goods exchange."

SERVICE ENTRY CO-ORDINATION

Although safety provisions are generally met, the physical co-ordination of service entry points to buildings in northern Canada remains haphazard. The water fill point may spill freezing water onto the entrance landing, making it slippery, or onto the ladder rung used by the fuel delivery person to reach the fuel tank. The sewage pump-out point may be too close to the boiler make-up air vent for comfort. The snowplow may bury or demolish garbage receptacles as it lumbers past. The to and fro of tank trucks in the confined spaces between buildings is poorly considered, judging by the high incidence of accidents involving small children and large vehicles in remote communities. In some new buildings users discover that the prevailing wind sends chimney exhaust into the main fresh-air intake of the mechanical ventilation system. Often the air intake is choked with snow during and after a blizzard, cutting off fresh air. In airtight buildings such "short circuits" and blockages have an immediate impact on the health of occupants. Each service enters the building at a point deemed most convenient to the preconceived interior plan. The lack of co-ordination arises partly from poor communication between the service bureaucracies and partly from lack of recognition that co-ordination of service delivery methods and details is crucial both to competent architecture and to a well-considered community plan. These pieces of the puzzle should interlock as surely as any other.

Service entry points imply penetration of the building envelope. Since most entry points are located in the lower sections of the building, where air infiltration induced by stack effect is greatest, the risk of freezing adjacent water pipes must be taken into account.

INTERIOR CO-ORDINATION

The space allowance in which to run pipes, ducts, and cable is restricted in any building, but scarcer still in compact buildings designed for high latitudes. For some designers the effort to resolve physical co-ordination problems is so great that many pipes or ducts finally run in spaces reserved for occupants.

There is no virtue in scrimping on the space legitimately required for mechanical rooms and heated crawl spaces. Constricted mechanical space slows the pace of maintenance work, which in turn obliges the owner to pay bigger labour bills. The location of mechanical rooms in the building plan is usually a compromise between a central position and one at the periphery that serves to shield the inhabited space from the winter wind. In the building proper, access panels convenient to user or maintainer must be co-ordinated thoroughly with mechanical services.

MAINTENANCE

In one northern Canadian building the initial operation and maintenance of the mechanical services was strangely inept for the first few months after commissioning, until an outsider discovered that the Number One (primary) boiler had been labelled Number Two (back-up). While proper labelling of machinery, pipes, direction of flow, and circuits is a standard requirement in building specifications, the installation work may be patchy. Examples such as this highlight the fact that if something can go wrong, it will go wrong. Keeping building services simple is too often overlooked as a design objective. Simplicity is a necessary antidote to the reality that few people in the world, much less in a remote community, are trained to trouble-shoot complex systems at short notice. In any case, complex systems that do not operate at full potential because of inadequate maintenance perform more primitively and more expensively than simple systems working well.

The standard specifications documents handed to maintenance personnel when a new building is commissioned are too dense and unco-ordinated to act as a guide for day-by-day operation. An additional document, or video documentary, edited by the engineer in plain language to explain the operation and interdependence of machines and controls would reduce the error and frustration felt by people introduced to complexities beyond their experience. Such a document would also serve incoming maintenance staff well.

AFTERWORD

The glass pyramid in the courtyard of the Louvre in Paris has become a symbol of the museum's resurgent popularity. It marks the new entrance to what used to be a rabbit warren of a palace housing national art treasures. It opens down to a huge orientation place below ground which neatly directs the visitor to any of the three main gallery wings. The frustrations of the warren have evaporated because visitors know where they stand at any moment and where they have go to meet the piece or period that interests them. Thanks to this pyramid and to three smaller skylights, the sky is always present in this space. All this makes the entrance to the Louvre a pleasure, a place where information and visitor can meet on friendly terms. The hugeness of the space is not overwhelming; there are many places to stop, lean on a broad railing, and gaze at the throng of humanity wending its way in and out of one of the world's great art museums. People flock here to enjoy themselves. Architectural detailing, choice of materials, and colours are laid-back in character but highly disciplined in execution. Nothing, no intersection of planes, no change of material, no fixture, happens by accident. The space works for humans because intelligence, understanding, imagination and discipline came together at one spot for one decade at the end of the second millenium. This is perhaps the best work of the Chinese-American architect I.M. Pei. The new Louvre is the perfect antidote to the modern architecture blues that most of us in North America endure day in and day out.

There is an old saying that architects are only as good as their clients. Disorganized owners, those who hardly know their own minds, let alone the precise requirements of their projects, can hire the best architects in the world and still come out losers. Such owners lose a golden opportunity, but more importantly, the people who have to use their buildings suffer for years to come. Too many people still believe that good building design is the stuff of high-flying architects and bottomless bank accounts. We have the Olympic Stadium in Montreal and the international airport in Denver—among countless other examples—to show that a failure of understanding at the start of a project inevitably leads to waste of time, good will and dollars. The hard part of a project is not the design, the construction, or the funding; it is right at the start where owners and their clients, the users of the building, have to come to grips with what is really meant by a project that fits their needs well. This is the programming or collective thinking stage, a period frequently deprived of time and resources in order to solve unrelated problems of the day. If the thinking is permitted to falter at this stage, everything that follows is doomed to miss the point. The building may stand, but it won't work. Let intelligence, understanding, imagination, and discipline intersect at the start of your building project, and see what happens!

In northern Canada a test case is developing. The Northwest Territories are subdividing, leaving the old capital and perhaps old ways of governing in the west, and forming Nunavut, with a new capital at Iqaluit in the east. Will this service town by the tidal flats transform itself into a thriving complex with positive meaning for every resident of Nunavut, or will it merely go on as before, dropping more unrelated buildings on the tundra, and hardening its attitude to the territory's farflung citizens? Nunavut has given itself an extraordinary opportunity, one whose outcome will set the tone for community development in the Canadian Arctic for the first part of the third millenium.

GLOSSARY

active layer That part of the soil that is above the permafrost table and that usually freezes in winter and thaws in summer. Also known as frost zone.

adfreeze bonding The process by which one object adheres to another by the binding action of ice.

albedo That fraction of the total light incident on a reflecting surface, especially a celestial body, that is reflected back in all directions.

Arctic Circle The parallel of latitude 66° 32' North.

ATV All-terrain vehicle.

aurora The most intense of the several lights emitted by the earth's upper atmosphere, seen most often along the outer realms of the Arctic and Antarctic, where it is called the aurora borealis and aurora australis respectively; excited by charged particles from space.

BP Years before present.

balloon framing A method of wood frame construction in which the bearing wall studs extend in one piece from the foundation wall to the eaves. The floor is framed into the studs.

boreal forest See *taiga*.

building envelope Those assemblies of construction materials that acting together separate the enclosed environment of a building from the outdoors; wall, floor, and roof assemblies, windows, skylights, and exterior doors.

BUR Built-up roofing, consisting of alternating layers of hot or solvent-based asphalt and roofing felts, covered by a thin layer of clean gravel laid in asphalt.

cathedral ceiling Sloped roof assembly, similar in construction and thickness to a wall assembly, whose ceiling finish is attached to the roof joists.

CFC Chlorofluorocarbon.

conventional roof assembly A roof assembly, usually sloped, in which main waterproofing and insulation are placed over the air/vapour barrier. Contrast *inverted roof assembly*.

crampons A pair of steel grids with spikes that fit over the soles of boots; held in place by straps; used for walking over moderately sloping ice or snow, and for steep ice or snow climbing.

cricket A low profile roof structure at the upstream junction of a large roof penetration and the plane of the roof to divert water around the penetration.

deadman A log, beam, or concrete block buried in the earth to anchor a wire rope attached to structural members above grade.

Dene Northern Athapaskan-speaking peoples of the Mackenzie River Valley and the Barren Grounds in the Northwest Territories.

dew point The temperature at which air becomes saturated when cooled without addition of moisture or change of pressure; any further cooling causes condensation. Also known as dew-point temperature.

drift All glacial and fluvio-glacial deposits left after the retreat of glaciers and ice sheets.

drumlin A hill of glacial *drift* or bedrock having a half-ellipsoidal streamline form like the inverted bowl of a spoon, with its long axis paralleling the direction of movement of the glacier that fashioned it.

drunken forest Trees leaning precariously because nearly uprooted by sloughing of the ground due to permafrost degradation.

EPDM Ethylene propylene diene monomer, a synthetic rubber used in the manufacture of some single-ply roof membrane systems. Early formulations highly susceptible to degradation when exposed to sunlight and/or low-level ozone. Incompatible with asphalt.

erratic A rock fragment that has been transported a great distance, generally by glacier ice or floating ice, and that differs from the bedrock on which it rests.

esker A sinuous ridge of constructional form, consisting of stratified accumulations, glacial sand, and gravel.

fast ice Sea ice generally remaining in the position where originally formed and sometimes attaining a considerable thickness; it is attached to the shore or over shoals, where it may be held in position by islands, grounded icebergs, or polar ice.

freezeback The freezing of soil with high moisture content to the surface of newly installed pile foundation members, to achieve adfreeze bonding.

frost heave The lifting and distortion of a surface due to internal action of frost resulting from subsurface ice formation; affects soil, rock, pavement, and other structures.

frost-susceptible soil Any soil or rock that is susceptible to frost heave as it is frozen.

hoarfrost A deposit of interlocking ice crystals formed by direct sublimation on objects. Also known as white frost.

icecap A perennial cover of ice and snow on a flat land mass such as an Arctic island.

ice-rich permafrost Any soil or rock that contains sufficient excess ice to cause significant settlement under load once it has been thawed.

iglu General term meaning "house" in Inuktitut (not snow house per se).

Inuit Eskimo-Aleut speaking peoples of the tundra regions spanning eastern Siberia, Alaska, Canada, and Greenland.

inukshuk A cairn having the rough shape and scale of an erect human; used by Inuit as a landmark or guide.

Inuktitut The language spoken by modern Inuit in several regional dialects.

inverted roof assembly A roof assembly, nearly horizontal, in which ballast, vapour-permeable protective sheet, and water-resistant insulation are installed above a main membrane that acts both as waterproofing and air/vapour barrier. Contrast *conventional roof assembly*.

isostatic rebound Over a period of thousands of years the rise of parts of the earth's crust by several tens of metres in response to the removal of the dead weight of the continental ice sheets.

komatik Sled (Inuktitut).

looming A form of mirage in which images of objects normally hidden below the horizon are seen in the sky, sometimes upside down; a common occurrence in the Arctic and Antarctic.

matchbox houses Slang for the very small wood frame houses erected in the Arctic by the Canadian government to shelter local populations during the 1950s and 1960s.

MBM Modified bitumen membrane; rubberized, reinforced asphalt sheets used as roofing membrane. Compatible with asphalt.

moisture content The quantity of water in a mass of soil, wood, or other material; expressed in percentage by weight of water in the mass.

moraine An accumulation of glacial *drift* deposited chiefly by direct glacial action and possessing initial constructional form independent of the floor beneath it.

muskeg A peat bog or tussock meadow, with variably woody vegetation.

nunatak An isolated hill, knob, ridge, or peak of bedrock projecting prominently above the surface of a glacier and completely surrounded by glacial ice.

patterned ground Any of several well-defined, generally symmetrical land forms, such as sorted and unsorted circles, polygons, and steps, that are characteristic of surficial material subject to intensive frost action.

pattern staining On interior painted surfaces of exterior walls and cathedral ceilings the pattern of the wood framing may "telegraph" through the paint surface over time, because of accelerated accretion of dust where the thermal differential through the wall is steepest.

PCB Polychlorinatedbiphenyl.

permafrost Perennially frozen ground, occurring wherever the temperature remains below 0°C for several years, whether the ground is actually consolidated by ice or not and regardless of the nature of the rock and soil particles of which the earth is composed.

permafrost table The upper limit of permafrost.

pingo A frost mound resembling a volcano, being a relatively large and conical mound of soil-covered ice, elevated by hydrostatic pressure of water within or below the permafrost.

platform framing A method of wood frame construction in which floor joists of each storey rest on the top plates of the bearing studs below or on the foundation wall. The floor is framed over the studs and completed to subflooring level to provide a platform for subsequent work.

polynya A Russian term for an open water area, other than a lead, lane, or crack, that is surrounded by sea ice.

polypan A lightweight pan for airproofing electrical connection boxes located in exterior wall assemblies (or acoustic partitions); made of thin plastic, it has flanges for lapping over air and/or vapour barrier membranes.

program A document prepared by or for the owner to instruct the architect in the general and detailed requirements of the building to be designed.

PVC Polyvinylchloride.

relative humidity The (dimensionless) ratio of the actual amount of water vapour present in a given parcel of air to the maximum amount of water vapour that parcel is capable of holding at a given temperature and pressure. Abbreviated rh.

rh Relative humidity.

roche moutonnée A small, elongate hillock of bedrock sculptured by a large glacier so that its long axis is oriented in the direction of ice movement; the upstream side is gently inclined, smoothly rounded, but striated, and the downstream side is steep, rough, and chipped.

RSI Thermal resistance of a material expressed as a Système International unit; divide by 0.176 for result in Imperial units of thermal resistance (R). (e.g., RSI 7.0 = R 40).

sastrugi Small ridges on the surface of snow, caused by winds and varying in size according to the strength and duration of the wind and the condition of the snow surface.

scree The accumulation of coarse rock material, typically at the bottom of a cliff or steep slope, due to mechanical weathering (e.g., frost action) of rocks and bedrock surfaces higher up.

sealift Common term for the annual resupply of remote communities by ship or barge.

sik-sik Arctic ground squirrel.

solifluction A rapid soil creep, especially referring to downslope movement in periglacial areas.

specific heat (capacity) Heat capacity of a substance divided by the mass; the quantity of heat required to raise the temperature of the unit mass of a substance by one degree.

stack effect The pronounced rising of heated air particularly in chimneys and multi-storey buildings due to the buoyancy of warmer air compared to ambient cooler air.

subarctic Pertaining to regions adjacent to the Arctic Circle or having characteristics somewhat similar to those of these regions.

sublimation The direct transition from solid to vapour, or vice versa, without any liquid phase being involved.

sull sash Secondary pane of glass with thin metal or wood edging fixed to the exterior of an insulating window or window frame with clips and screws.

sundog Parhelion; either of two coloured luminous spots at 22° (or somewhat more) on both sides of the sun and at the same elevation as the sun, as seen from the ground.

taiga A zone of forest vegetation encircling the Northern Hemisphere between the arctic-subarctic tundra in the north and the steppes, hardwood forests, and prairies in the south. Also *boreal forest*.

thaw-stable permafrost Any soil or rock that contains no excess ice and that will not settle under load once it has been thawed.

thermal break Any material or air gap that acts to reduce the potential heat loss posed by the presence of a thermal bridge in a building envelope.

thermal bridge Any material having thermal conductivity much higher than that of adjacent materials and positioned in a building envelope such that it bridges all or much of the distance between the heated interior and the unheated exterior, thereby permitting more heat per unit area to be conducted to the outside compared to the other materials in the same assembly.

thermokarst topography An irregular land surface formed in a permafrost region by melting ground ice.

till Unsorted and unstratified *drift* consisting of a heterogeneous mixture of clay, sand, gravel, and boulders that is deposited by and underneath glaciers and ice sheets.

tree line The atitudinal or latitudinal limits for arboreal growth.

tundra The treeless plains of northern North America and northern Eurasia, lying principally along the Arctic Circle, and on the northern side of the taiga.

tussock A small hummock of generally solid ground in a muskeg or marsh, usually covered with and bound together by the roots of low vegetation such as grasses, sedges, or shrubs of the heath family.

utilidor An insulated, heated conduit built below the ground surface or supported above the ground surface to protect the contained water, steam, sewage, and fire protection lines from freezing.

utilidette A small utilidor incorporated in or just below the floor structure of a group of attached buildings such as row houses.

vapour pressure The partial pressure of water vapour in the atmosphere.

whiteout An atmospheric optical phenomenon of the polar regions in which the observer appears to be engulfed in a uniformly white glow; shadows, horizon, and clouds are not discernible. Sense of depth and orientation are lost; dark objects in the field of view appear to float at an indeterminable distance.

BIBLIOGRAPHY

PART I - PEOPLE

Abele, F. (1989). *Gathering strength.* Calgary:Arctic Institute of North America.

Aigner, J.S. (1985). "Early arctic settlements in North America." *Scientific American,* 253(5).

Arnaktauyok, G. (1976). *Stories from Pangnirtung.* Edmonton: Hurtig.

Berger, T.R. (1985). *Village journey: The report of the Alaska Native Review Commission.* New York: Hill & Wang.

Blondin, G. (1990). *When the world was new: Stories of the Sahtu Dene.* Yellowknife: Outcrop.

Bruemmer, F. (1971). *Seasons of the Eskimo: A vanishing way of life.* Toronto: McClelland & Stewart.

Canadian Arctic Resources Committee. (1987). *Boundary and constitutional agreement for the implementation of division of the Northwest Territories.* Iqaluit.

Chatwin, B. (1987). *The Songlines.* London: Picador-Pan.

Condon, R.G. et al. (1995). "The best part of life: Subsistence hunting, ethnicity, and economic adaptation among young Inuit males." Calgary: *Arctic,* 48(1).

Coté, C.P., and Dufour, J. (1984). *Le nord du Québec: Profil régional.* Québec: Gouvernement du Québec.

Crnkovich, M. (1990). *Gossip: A spoken history of women in the North.* Ottawa: Canadian Arctic Resources Committee.

Crowe, K. (1974). *A history of the original peoples of Northern Canada.* Montreal: Arctic Institute of North America.

Cserepy, F. (1984). *Native languages in the Northwest Territories.* Yellowknife: Government of the NWT.

Department of Education, GNWT. (1978). *Philosophy of education in the Northwest Territories.* Yellowknife.

Dene Cultural Institute. (1989). *Decho.* Yellowknife.

Documentation Photographique. (1982). *L'homme dans les pays froids arctiques et subarctiques.* Paris: La Documentation Française.

Dolitsky, A.B. (1985). "A critical review of 'the Mesolithic' in relation to Siberian archaeology." *Arctic,* 38(3).

Eber, D.H. (1989). *When the whalers were up north.* Boston: Godine.

Goodman, J. (1981). *American genesis.* New York: Summit.

Gould, S.J. (1989). *Wonderful life: The Burgess Shale and the nature of history.* New York: Norton.

Government of Canada. (1972). *Interim report of northern housing.* Ottawa: Public Works.

Government of Denmark. (1983). *A town in Greenland.* Copenhagen.

Government of the NWT. (1989). *Statistics Quarterly.* Yellowknife.

———. (1983). *Residential user needs study for Lake Harbour, NWT.* Yellowknife: Department of Local Government.

Grant, Shelagh D. (1991). "A case of compounded error: The Inuit resettlement project, 1953, and the Government response, 1990." *Northern Perspectives,* 19(1).

Hall, S. (1988). *The fourth world: The heritage of the Arctic and its destruction.* New York: Vintage.

Helmer, J.W. (1991). "The Paleo-Eskimo history of the North Devon Lowlands." *Arctic,* 44(4).

Herbert, M. (1973). *The snow people.* London: Book Club.

Honigman, J.J. and I. (1965). *Eskimo townsmen.* Ottawa: University of Ottawa.

Houston, J. (1995) *Confessions of an igloo dweller.* Toronto: McClelland & Stewart.

Innukshuk, R. and Cowan, S. (1976). *We don't live in snow houses anymore.* Ottawa: Canadian Arctic Producers.

Irwin, C. (1989). *Lords of the Arctic: Wards of the state.* Halifax: Dalhousie University.

Janes, R.R. (1976). "Culture contact in the 19th-century Mackenzie Basin, Canada." *Current Anthropology,* 17(2).

———. (1982). "The preservation and ethnohistory of a frozen historic site in the Canadian Arctic." *Arctic,* 35(3).

———. (1983). *Archaeological ethnography among Mackenzie Basin Dene, Canada.* Calgary: Arctic Institute of North America.

Jenness, A. and Rivers, A. (1989). *In two worlds: A Yup'ik Eskimo family.* Boston: Houghton Mifflin.

Leakey, R.E. (1981). *The making of mankind.* New York: Dutton.

Levine, L.D., ed. (1975). *Man in nature: Historical*

perspectives on man in his environment. Toronto: Royal Ontario Museum.

Lewontin, R. (1982). *Human diversity.* New York: Scientific American Books.

Lipke, W.C. (1987). *Inuit art: Contemporary perspectives.* Washington: American Review of Canadian Studies.

Lyall, E. (1979). *An arctic man.* Edmonton: Hurtig.

MacInnis, J. (1982). *The Breadalbane adventure.* Montreal: Optimum.

McCullough, K.M. (1989). *The Ruin Islanders.* Ottawa: Canadian Museum of Civilization.

McGhee, R. (1978). *Canadian arctic prehistory.* Toronto: Van Nostrand Reinhold.

——. (1989). *Ancient Canada.* Ottawa: Canadian Museum of Civilization.

McMillan, A.D. (1988). *Native peoples and cultures of Canada.* Vancouver: Douglas & McIntyre.

Matthiasson, J.S. (1992). *Living on the land: Change among the Inuit of Baffin Island.* Peterborough: Broadview.

Matthiessen, P. (1984). *Indian country.* New York: Viking.

——. (1991). *In the spirit of Crazy Horse.* New York: Viking.

Milligan, S. (1986). *Living explorers of the Canadian Arctic.* Yellowknife: Outcrop.

Morrison, D.A. (1988). *The Kugaluk site and the Nuvorugmiut: The archaeology and history of a nineteenth century Mackenzie Inuit society.* Ottawa: Canadian Museum of Civilization.

Mowat, F. (1973). *Sibir: My discovery of Siberia.* Toronto: McClelland & Stewart.

——. (1973). *Tundra.* Toronto: McClelland & Stewart.

Nungaq, Anna, Quoted at Grise Fiord. *News/North,* 6 June 1988. [Yellowknife]

Outcrop Ltd. (1990). *Northwest Territories Data Book 1990-91.* Yellowknife.

Pryde, D. (1972). *Nunaga: Ten years among the Eskimo.* London: Eland.

Richardson, B. (1993). *People of Terra Nullius.* Vancouver: Douglas & McIntyre.

Rogers, E.S. (1970). *Indians of the subarctic.* Toronto: Royal Ontario Museum.

Steltzer, U. (1982). *Inuit: The North in transition.* Vancouver: Douglas & McIntyre.

Vezinet, M. (1980). *Les Nunamiut: Inuit au coeur des terres.* Québec: Ministère des affaires culturelles.

Weeden, R.B. (1985). "Northern people, northern resources, and the dynamics of carrying capacity." *Arctic,* 38(2).

Yukon Territorial Government. (1984). *Yukon data book 1984-85.* Yellowknife: Outcrop.

PART II - TERRAIN

The Alaska-Yukon wild flowers guide. 1974. Anchorage: Alaska Magazine.

Bliss, L.C., ed. (1987). *Truelove Lowland, Devon Island, Canada: A high arctic ecosystem.* Edmonton: University of Alberta Press.

Bone, R.M. (1992). *The geography of the Canadian North: Issues and challenges.* Toronto: Oxford University Press.

Briggs, C.D. (1979). *Plant magic for northern gardens.* Yellowknife: Northern News Service.

Brown, R.J.E. (1983). *Effects of fire on the permafrost ground thermal regime.* Ottawa: National Research Council of Canada.

Burt, P. (1991). *Barrenland beauties: Showy plants of the arctic coast.* Yellowknife: Outcrop.

Calef, G. (1981). *Caribou and the Barren-lands.* Ottawa: Canadian Arctic Resources Committee.

Davis, R.A. et al. (1980). *Arctic marine mammals in Canada.* Yellowknife: Science Advisory Board of the Northwest Territories.

Edlund, S.A. (1987). "Plants: Living weather stations." In *Geos.* Ottawa: Energy Mines and Resources Canada.

——, and Egginton, P.A. (1984). "Morphology and description of an outlier population of tree-sized willows on western Victoria Island, NWT." In *Current Research.* Ottawa: Geological Survey of Canada.

Eides Forlag, J.W., pub. (1961). *Spitzbergen.* Norway.

Goodrich, L.E. (1982). *The influence of snow cover on the ground thermal regime.* Ottawa: National Research Council of Canada.

Graves, J. and Hall, E. (1985). *Arctic animals.* Yellowknife: Renewable Resources, GNWT.

Hall, E., ed. (1989). *People and caribou in the Northwest Territories.* Yellowknife: Government of the Northwest Territories.

Hoyt, E. (1984). *The whales of Canada.* Camden East: Camden House.

Lopez, B. (1986). *Arctic dreams.* New York: Scribner's.

McLaren, M.A. and Green, J.E. (1985). "The reactions of muskoxen to snowmobile harassment." *Arctic,* 38(3).

Melzack, R. (1970). *Raven: Creator of the world.* Boston: Little Brown.

Muller, F. (1977). *The living Arctic.* Toronto: Methuen.

Porsild, A.E. and Cody, W.J. (1980). *Vascular plants of continental Northwest Territories.* Ottawa: National Museums of Canada.

Ray, G.C. and McCormick-Ray, M.G. (1981). *Wildlife of the polar regions.* New York: Abrams.

Sater, J.E., ed. (1969). *The arctic basin.* Washington: Arctic Institute of North America.

Smith, W.T. and Cameron, R.D. (1985). "Reactions of large groups of caribou to a pipeline corridor on the arctic coastal plain of Alaska." *Arctic,* 38(1).

Strub, H. (1992). "Swim by an Arctic Fox, Alopex lagopus, in Alexandra Fiord, Ellesmere Island, Northwest Territories." *Canadian Field-Naturalist,* 106(4).

Svoboda, J. and Freedman, B., ed. (1994). *Ecology of a polar oasis: Alexandra Fiord, Ellesmere Island, Canada.* Toronto: Captus University Publications.

Symington, F. (1965). *Tuktu: A question of survival.* Ottawa: Canadian Wildlife Service.

Walker, M. (1984). *Harvesting the northern wild.* Yellowknife: Outcrop.

Wiebe, R. (1989). *Playing dead: A contemplation concerning the Arctic.* Edmonton: NeWest.

Young, S.B. (1989). *To the Arctic: An introduction to the far northern world.* New York: Wiley.

PART III - CLIMATE

Atmospheric Environment Service (1982-84, 1990). *Canadian climate normals: Vols. 1-9; Solar radiation, Temperature, Precipitation, Degree days, Wind, Frost, Bright sunshine, Pressure Temperature Humidity, Soil temperature, Lake evaporation, Days with ...,.* Ottawa: Environment Canada.

———. (1982). *Canadian climate normals: Temperature and precipitation: The North – YT and NWT.* Ottawa: Environment Canada.

Culjat, B. (1975). *Climate and the built environment in the North.* Thesis. Stockholm: Arkitektursektionens tryckeri, KTH.

Lohvinenko, T.W. and Zalcik, M.S. (1991). "Noctilucent clouds seen from North America." *Arctic,* 44(4).

Matus, V. (1988). *Design for northern climates.* New York: Van Nostrand Reinhold.

Osczevski, R.J. (1995). "The basis of wind chill." *Arctic,* 48(4).

Schaefer, V.J. and Day, J.A. (1981). *A field guide to the atmosphere.* Boston: Houghton Mifflin.

Shah, G.M. and Wardle, D.I. (1986). *One year's data on UVB spectral irradiance at Toronto.* Toronto: Atmospheric Environment Service.

Thomas, M.K. (1961). "A survey of temperatures in the Canadian Arctic." In *Geology of the Arctic.* Toronto: University of Toronto.

———, and Boyd, D.W. (1957). "Wind chill in northern Canada." *The Canadian Geographer,* 10.

Thompson, H.A. (1965). *The climate of the Canadian Arctic.* Ottawa: Transport Canada.

———. (1965). *The climate of Hudson Bay.* Ottawa: Energy Mines and Resources Canada.

Williams, C.J. (1984). *Arctic snowdrifting.* Guelph, ON: Morrison Hershfield.

PART IV - PROGRAM

Aulicems, A. et al. (1973). "Winter clothing requirements for Canada." *Climatological Studies,* 22. Toronto: Environment Canada.

Barr, W. and Wilson, E.A. (1985). "The shipping crisis in the Soviet eastern Arctic at the close of the 1983 shipping season." *Arctic,* 38(1).

Bassett, M.R. et al. (1981). *An appraisal of the sulphur hexafluoride decay technique for measuring air infiltration rates in buildings.* Ottawa: National Research Council of Canada.

Beattie, O. and Geiger, J. (1987). *Frozen in time.* Saskatoon: Prairie Books.

Bent, C. (1983). "Mobility and flexibility in northern communities." *Environments,* 15(2).

Berglund, B. (1974). Wilderness survival. Toronto: Pagurian Press.

Blake, P. (1977). *Form follows fiasco: Why modern architecture has not worked.* Boston: Atlantic/Little, Brown.

Brolin, B.C. (1976). *The failure of modern architecture.* New York: Van Nostrand Reinhold.

Canada Mortgage and Housing Corporation (1984). *Rural and native housing 1984. Client survey summary report.* Ottawa.

———. (1987). *Examples of housing construction in the North.* Ottawa.

Canadian Transport Commission (1979). *Cargo utilization: Selected aspects of pallets, barges, containers and van trailers in multi-modal transport.* Ottawa: Government of Canada.

Carlsson, B. et al. (1980). *Airtightness and thermal insulation: Building design solutions.* Stockholm: Swedish Council for Building Research.

Caudill, W.W. et al. (1981). *Architecture and you: How to experience and enjoy buildings.* New York: Whitney Library of Design.

Chill, R.W. and Latta, J.K. (1985). *The unforgiving north.* Ottawa: National Research Council of Canada.

Dawson, P.C. (1995). "'Unsympathetic users': An ethnoarchaeological examination of Inuit responses to the changing nature of the built environment." *Arctic,* 48(1).

Drew, P. (1979). *Tensile architecture.* London: Granada.

Erskine, R. (1964). The challenge of the high latitudes, and Community design for production, for publication or for the people? *RAIC Journal*, 41(1).

———. (1966). "Architecture and town planning in the north." *Polar Record*, 14(89).

Fenge, T. and Rees, W.E., eds. (1987). *Hinterland or homeland?: Land-use planning in northern Canada.* Ottawa: Canadian Arctic Resources Committee.

Friedman, Y. (1980). *Toward a scientific architecture.* Cambridge, MA: MIT Press.

Government of the NWT. (1983). *Residential user needs study for Lake Harbour, NWT.* Yellowknife: Department of Local Government, GNWT.

Hall, E.T. (1977). *Beyond culture.* New York: Doubleday Anchor Press.

Hall, P. (1980). *Great planning disasters.* Los Angeles: University of California Press.

Hancock, L. (1988). *Gino Pin: Northern architect.* Up Here [Yellowknife].

Heimsath, C. (1977). *Behavioral architecture: Toward an accountable design process.* New York: McGraw-Hill.

Heinke, G.W. and Bowering, E.J. (1982). *Community fire protection and prevention north of 60.* Yellowknife: Governments of the NWT & Yukon.

Heschong, L. (1979). *Thermal delight in architecture.* Cambridge, MA: MIT Press.

Hiss, T. (1990). *The experience of place.* New York: Knopf.

Hutcheon, N.B. and Handegord, G.O. (1980). *Evolution of the insulated wood-frame house in Canada.* Ottawa: National Research Council of Canada.

Issenman, B. and Rankin, C. (1988). *Ivalu: Traditions of Inuit clothing.* Montreal: McCord Museum.

Jacobsen, G. (1968). "Canada's northern communities." In *North*. Ottawa: Indian and Northern Affairs.

Kalman, H. (1994). *History of Canadian architecture.* Don Mills, ON: Oxford University Press.

Lancaster Sound Regional Land Use Planning Commission. (1991). *The Lancaster Sound regional land use plan.* Yellowknife: Northwest Territories Land Use Planning.

Lysyk, K.M. et al. (1977). *Alaska Highway Pipeline Inquiry.* Ottawa: Supply and Services Canada.

McHarg, I.L. (1969). Design with nature. New York: Natural History Press.

Nabakov, P. and Easton, R. (1989). *Native American architecture.* New York: Oxford University Press.

National Research Council of Canada. (1983). *Measures for energy conservation in new buildings.* Ottawa.

Northwest Territories Housing Corporation. (1987). *Building houses, communities and our future.* Yellowknife.

Northwest Territories Legislative Assembly. (1984). *Interim report of the special committee on housing.* Yellowknife: Government of the Northwest Territories.

Nottingham, D. and Stevens, M. (1984). "An arctic house." *The Northern Engineer*, 16(1).

Ott, J.N. (1976). *Health and light.* New York: Pocket Books.

Pedersen, G.L. et al. (1980). *Journey to the Northwest Territories.* Copenhagen: Greenland Technical Organization.

Pressman, N. (1985). *Reshaping winter cities: Concepts, strategies and trends.* Waterloo: University of Waterloo.

Quirouette, R.L (1985). *The difference between a vapour barrier and an air barrier.* Ottawa: National Research Council of Canada.

Rasmussen, S.E. (1962). *Experiencing architecture.* Cambridge, MA: MIT Press.

Robinson, G. and Baker, M.C. (1975). *Wind-driven rain and buildings.* Ottawa: National Research Council of Canada.

Rudofsky, B. (1964). *Architecture without architects.* New York: Doubleday.

Schoenauer, N. (1973). *Introduction to contemporary indigenous housing.* Montreal: Reporter Books.

Schuyler, G.D. and Solvason, K.R. (1983). *Effectiveness of wall insulation.* Ottawa: National Research Council of Canada.

Scott, D.L. (1983). *Seasonal moisture variation in wood siding and framing: A case study.* Ottawa: National Research Council of Canada.

Sommer, R. (1969). *Personal space: The behavioral basis of design.* Englewood Cliffs, NJ: Prentice-Hall.

Steltzer, U. (1981). *Building an igloo.* Vancouver: Douglas & McIntyre.

Strub, H. (1986). *Kivalliq Hall post-occupancy evaluation – Rankin Inlet, NWT.* Yellowknife: Public Works and Highways, GNWT.

———. (1971). *Arctic habitation.* Thesis, University of British Columbia, Vancouver.

Underground Space Center. (1979). *Earth sheltered housing design.* New York: University of Minnesota, Van Nostrand Reinhold.

Wolfe, T. (1981). *From Bauhaus to our house.* New York: Farrar Strauss Giroux.

Yue, D. and C. (1988). *The igloo.* Boston: Houghton Mifflin.

PART V - DESIGN

Allen, E. (1990). *Fundamentals of building construc-*

tion: Materials and methods. New York: Wiley.

Baker, M.C. (1980). *Roofs.* Montreal: Multi-Science Publications.

Baker, M.C. (1984). *New roofing materials.* Ottawa: National Research Council of Canada.

Brand, R. (1990). *Architectural details for insulated buildings.* New York: Van Nostrand Reinhold.

Burn, K.N. and Roux, R. (1982). *Insulating existing flat roofs: Design and construction details.* Ottawa: National Research Council of Canada.

Canada Mortgage and Housing Corporation. (1977). *The conservation of energy in housing.* Ottawa.

———. (1979). *Roof decks design guidelines.* Ottawa.

———. (1981). *Log house construction requirements.* Ottawa.

———. (1982). *Energy-efficient housing construction.* Ottawa.

———. (1982). *A glossary of house-building and site-development terms.* Ottawa.

———. (1983). *Space frame foundation system for permafrost.* Ottawa.

———. (1984). *Canadian wood-frame house construction.* Ottawa.

———. (1984). *Rural and native housing client survey summary report.* Ottawa.

———. (1984). *Trouble-free windows, doors and skylights.* Ottawa.

———. (1984). *Ventilation for humidity control.* Ottawa.

———. (1985). *Seminar: Heating systems for arctic and subarctic Canada.* Ottawa.

———. (1989). *Moisture and air: Problems and remedies.* Ottawa.

Carlson, A.R. (1983). *Log versus frame house.* Typescript. Fairbanks: University of Alaska.

———. (1977). *Building a log house in Alaska.* Fairbanks: Cooperative Extension Service, University of Alaska.

———. (1976-80). *Building in Alaska.* [Pamphlet series]. Fairbanks: Cooperative Extension Service, University of Alaska.

Collymore, P. (1982). *The architecture of Ralph Erskine.* London: Granada.

Crittenden, E.B. (1988). "Cold dry climate construction." In *Encyclopedia of architecture: Design, engineering and construction.* New York: Wiley.

Egelius, M. (1982). "Ralph Erskine: An architect on the move." In *Space and Society,* 17. Milan: Sansoni/MIT Press.

EPEC Consulting Ltd. (1981). *Community water and sanitation services: Northwest Territories.* Yellowknife: Government of the NWT.

Erskine, R. (1973). *Diary of study journey to Greenland.* Unpublished. Drottningholm, Sweden.

———, and Culjat, B. (1974). *Resolute Bay New Town.* Project summary. Stockholm: Ralph Erskine Architect and Planner.

Evans, B.H. (1981). *Daylight in architecture.* New York: McGraw-Hill.

Forgues, Y.E. (1985). *The ventilation of insulated roofs.* Ottawa: National Research Council of Canada.

Givoni, B. (1981). *Man, climate and architecture.* New York: Van Nostrand Reinhold.

Glover, M., ed. (1974). *Building in northern communities.* Montreal: Arctic Institute of North America.

Greenland Technical Organization. (1984). *Guidelines for housing and human settlement planning and design criteria in cold climates: Greenland.* Copenhagen.

Halse, A.O. (1978). *The use of color in interiors.* New York: McGraw-Hill.

Hough Stansbury Woodland Ltd. (1991). *Winter Cities Design Manual.* City of Sault Ste. Marie.

Housing and Urban Development. (1982). *An analysis of the feasibility of utilizing factory-built and other appropriate types of housing for Indians and Alaskan Natives.* Washington.

Hutcheon, N.B. and Handegord, G.O. (1983). *Building science for a cold climate.* Toronto: Wiley.

Jensen, T. (1983). *Skylights.* Philadelphia: Running Press.

Laaly, H.O. (1982). *Methods of evaluating single-ply roofing membranes.* Ottawa: National Research Council of Canada.

Lam, W.M. (1977). *Perception and lighting as form-givers for architecture.* New York: McGraw-Hill.

Larsson, N.K. (1985). *Energy considerations in the design of northern housing.* Ottawa: National Research Council of Canada.

Latta, J.K. (1973). *Walls, windows and roofs for the Canadian climate.* Ottawa: National Research Council of Canada.

———. (1985). *The principles and dilemmas of designing durable house envelopes for the North.* Ottawa: National Research Council of Canada.

Magid, H. (1983). "Shelter design in remote settlements: an examination of Inuit shelter design experiences in the Canadian Arctic." Waterloo: *Environments,* 15(2).

Manty, J. and Pressman, N. (1988). *Cities designed for winter.* Tampere: Building Book.

Moore, C., and Allen, G. (1976). *Dimensions: Space, shape and scale in architecture.* New York: Architectural Record Books.

National Research Council of Canada. (1976). *Cracks, movements and joints in buildings.* Ottawa.

National Research Council of Canada. (1980). *Construction details for airtightness.* Ottawa.

——. (1988). *Window performance and new technology.* Ottawa.

——. (1995). *National Building Code of Canada.* Ottawa.

Norman, D. (1990). *The design of everyday things.* New York: Doubleday Currency.

Pearson, P.D. (1978). *Alvar Aalto and the International Style.* New York: Whitney Library of Design.

Peterson, I. (1995). "Viewing frost heave on a microscopic scale." *Science News,* 148(1).

Porter, T. (1982). *Architectural color: A design guide to using color on buildings.* New York: Whitney Library of Design.

Shurcliff, W.A. (1980). *Thermal shutters and shades.* Andover, MA: Brick House.

Smith, B. (1984). *Introduction to foundation engineering in northern Canada.* Calgary: Thurber Consultants Ltd.

Strub, H. (1983). Log construction feasibility study: Mackenzie Valley. Draft report. Yellowknife: Department of Public Works, GNWT.

Taylor, D.A. (1983). *Sliding snow on sloping roofs.* Ottawa: National Research Council of Canada.

Templin, J.T. and Schriever, W.R. (1982). *Loads due to drifted snow.* Ottawa: National Research Council of Canada.

van Ginkel Associates Ltd. (1976). *Building in the North: Responding to the environment.* Canadian Arctic Gas Study Ltd.

——. (1976). *Building in the North: Experience and projects.* Canadian Arctic Gas Study Ltd.

APPENDIX A

CLIMATE CHARTS

The data contained in the following charts were obtained scientifically by the Atmospheric Environment Service of Canada from a network of climate observing stations. Combining the data as shown, especially putting up the average results from northern Canadian communities against those of southern Canadian cities, is not scientific: it is impressionistic. However, since most building professionals working in northern Canada were trained in southern Canada, the comparison may be useful in assisting designers and builders to understand the effects of the northern Canadian climate on high latitude building design.

1.31	SOLAR ALTITUDE AT NOON	165
2.31	DAYLIGHT HOURS	166
3.31	DAYLIGHT	167
4.31	TOTAL BRIGHT SUNSHINE	168
5.31	MEAN DAILY TEMPERATURE	169
6.31	SOIL TEMPERATURE	170
7.31	NUMBER OF DAYS WITH FROST	171
8.31	DEGREE-DAYS BELOW 18°C	172
9.31	MEAN WIND SPEED	173
10.31	PREVAILING WIND DIRECTION	174
11.31	MEAN WIND CHILL COOLING RATE	175
12.31	TOTAL PRECIPITATION	176
13.31	MEAN SNOWFALL	177
14.31	NUMBER OF DAYS WITH BLOWING SNOW	178
15.31	RELATIVE HUMIDITY	179
16.31	VAPOUR PRESSURE	180
17.31	NUMBER OF DAYS WITH FOG	181
18.31	DESIGN DATA FOR EIGHT NORTHERN AND EIGHT SOUTHERN CANADIAN CENTRES	182

1.31 SOLAR ALTITUDE AT NOON Degrees of arc, 3rd week of the month

Lat	Location	JAN	FEB	MAR	APR	MAY	JUN	JUL	AUG	SEP	OCT	NOV	DEC	Year
60	Whitehorse	10	18	29	40	50	53	50	40	29	18	10	6	29
68	Inuvik	2	11	22	33	43	45	43	33	22	11	2	0	22
61	Fort Simpson	9	17	28	39	49	52	49	39	23	17	9	5	28
62	Yellowknife	8	17	28	39	48	51	48	39	28	17	8	4	28
69	Cambridge Bay	2	10	21	32	42	44	42	32	21	10	2	0	22
64	Baker Lake	6	15	26	37	47	49	47	37	26	15	6	2	26
74	Resolute	0	4	15	26	36	39	36	26	15	4	0	0	17
63	Iqaluit	7	15	26	37	47	50	47	37	26	15	7	3	26
	North MEAN	**6**	**13**	**24**	**35**	**45**	**48**	**45**	**35**	**24**	**13**	**6**	**3**	**25**
49	Vancouver	22	30	41	52	62	65	62	52	41	30	22	18	41
53	Edmonton	18	26	37	48	58	61	58	48	37	26	18	14	37
50	Regina	21	29	40	51	61	64	61	51	40	29	21	17	40
50	Winnipeg	21	29	40	51	61	64	61	51	40	29	21	17	40
46	Sudbury	25	33	44	55	65	68	65	55	44	33	25	21	44
43	Toronto	28	36	47	58	68	69	68	58	47	36	28	24	47
45	Montreal	26	34	45	56	66	69	66	56	45	34	26	22	45
44	Halifax	27	35	46	57	67	70	67	57	46	35	27	23	46
	South MEAN	**24**	**32**	**43**	**54**	**64**	**66**	**64**	**54**	**43**	**32**	**24**	**20**	**43**

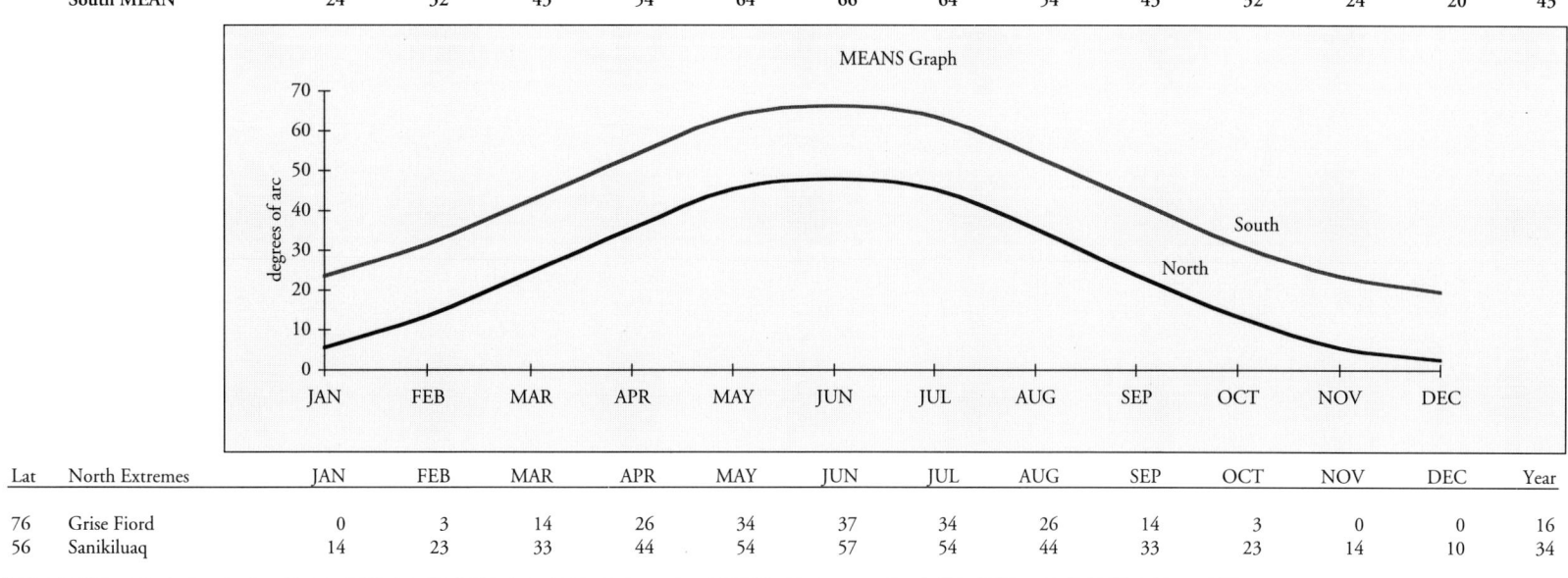

MEANS Graph

Lat	North Extremes	JAN	FEB	MAR	APR	MAY	JUN	JUL	AUG	SEP	OCT	NOV	DEC	Year
76	Grise Fiord	0	3	14	26	34	37	34	26	14	3	0	0	16
56	Sanikiluaq	14	23	33	44	54	57	54	44	33	23	14	10	34

APPENDIX A

2.31 DAYLIGHT HOURS   Hours per day at midmonth

Lat	Location	JAN	FEB	MAR	APR	MAY	JUN	JUL	AUG	SEP	OCT	NOV	DEC	Year
60	Whitehorse	6.7	9.2	11.7	14.6	17.1	18.8	18.0	15.7	12.9	10.0	7.5	5.9	12.3
68	Inuvik	0.7	7.8	11.6	15.8	21.0	24.0	24.0	17.6	13.4	9.4	4.7	0.2	12.5
61	Fort Simpson	6.4	9.1	11.7	14.7	17.4	19.4	18.5	15.9	12.9	10.1	7.3	5.5	12.4
62	Yellowknife	6.1	8.9	11.7	14.8	17.7	20.1	19.0	16.1	13.0	10.1	7.0	5.0	12.5
69	Cambridge Bay	0.5	7.5	11.6	16.0	21.8	24.0	24.0	18.0	13.4	9.2	4.3	0.1	12.5
64	Baker Lake	5.3	8.5	11.7	15.0	18.4	21.3	19.7	16.5	13.2	9.8	6.5	4.1	12.5
74	Resolute	0.0	5.5	11.4	17.6	24.0	24.0	24.0	23.1	13.8	8.2	0.0	0.0	12.6
63	Iqaluit	5.7	8.7	11.7	14.9	18.1	20.7	19.4	16.3	13.0	10.0	6.7	4.6	12.5
	North MEAN	**3.9**	**8.2**	**11.6**	**15.4**	**19.4**	**21.5**	**20.8**	**17.4**	**13.2**	**9.6**	**5.5**	**3.2**	**12.5**
49	Vancouver	8.7	10.2	11.8	13.8	15.3	16.1	15.7	14.5	12.6	10.7	9.2	8.3	12.2
53	Edmonton	7.9	9.8	11.8	14.1	15.9	17.1	16.5	14.9	12.7	10.5	8.6	7.4	12.3
50	Regina	8.5	10.1	11.8	13.8	15.4	16.4	15.9	14.5	12.7	10.7	9.1	8.1	12.3
50	Winnipeg	8.5	10.1	11.8	13.8	15.4	16.4	15.9	14.5	12.7	10.7	9.1	8.1	12.3
46	Sudbury	9.3	10.3	11.8	13.7	15.0	16.0	15.4	14.3	12.6	10.8	9.4	8.6	12.3
43	Toronto	9.3	10.5	11.8	13.4	14.6	15.2	15.0	14.1	12.5	11.0	9.8	9.1	12.2
45	Montreal	9.1	10.4	11.9	13.6	14.8	15.7	15.4	14.2	12.5	10.9	9.6	8.7	12.2
44	Halifax	9.2	10.5	11.9	13.5	14.7	15.5	15.1	14.2	12.5	10.9	9.7	8.9	12.2
	South MEAN	**8.8**	**10.2**	**11.8**	**13.7**	**15.1**	**16.1**	**15.6**	**14.4**	**12.6**	**10.8**	**9.3**	**8.4**	**12.2**

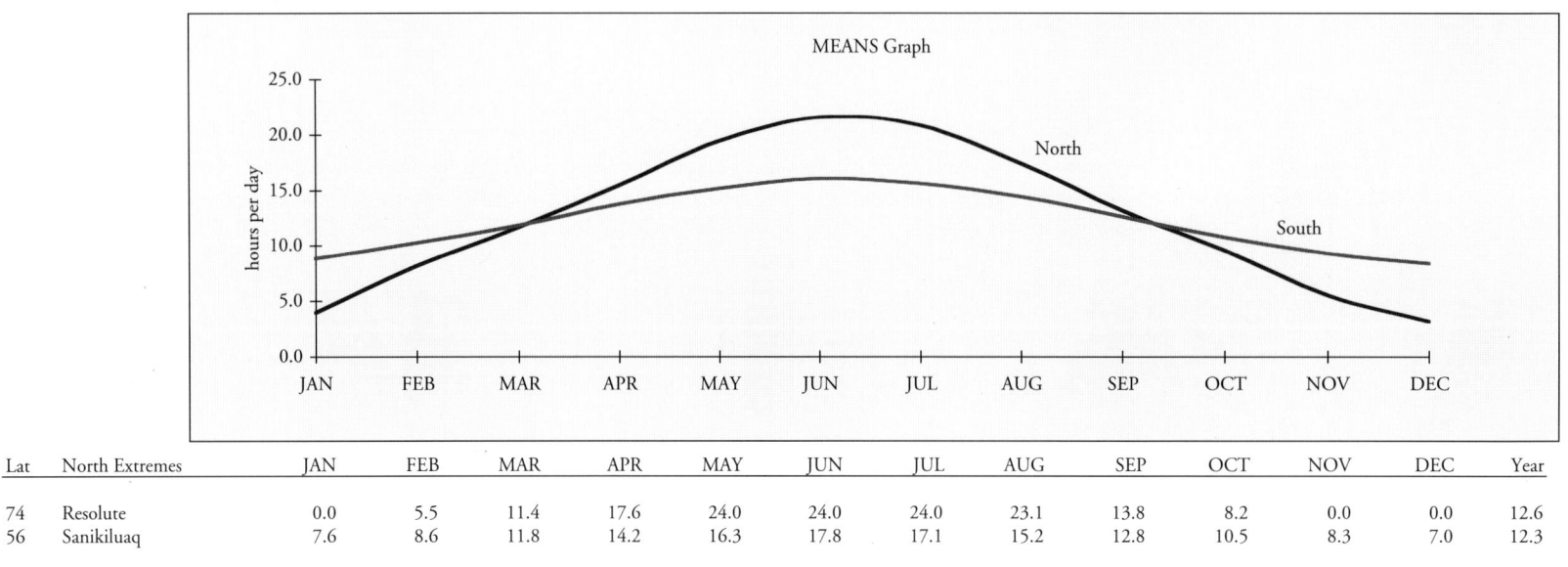

MEANS Graph

Lat	North Extremes	JAN	FEB	MAR	APR	MAY	JUN	JUL	AUG	SEP	OCT	NOV	DEC	Year
74	Resolute	0.0	5.5	11.4	17.6	24.0	24.0	24.0	23.1	13.8	8.2	0.0	0.0	12.6
56	Sanikiluaq	7.6	8.6	11.8	14.2	16.3	17.8	17.1	15.2	12.8	10.5	8.3	7.0	12.3

3.31 DAYLIGHT Hours per month, total possible sunshine

Lat	Location	JAN	FEB	MAR	APR	MAY	JUN	JUL	AUG	SEP	OCT	NOV	DEC	Year
60	Whitehorse	213	257	367	438	533	561	557	483	386	314	227	187	4523
68	Inuvik	99	218	364	474	647	720	705	539	397	287	149	13	4612
61	Fort Simpson	204	253	366	441	541	573	566	488	387	302	220	177	4518
62	Yellowknife	195	249	366	445	549	587	577	493	388	309	213	166	4537
69	Cambridge Bay	69	210	363	481	671	720	724	551	399	282	129	0	4599
64	Baker Lake	174	241	365	453	571	621	605	504	391	303	197	138	4563
74	Resolute	0	148	360	535	743	720	744	648	413	246	24	0	4581
63	Iqaluit	185	245	366	449	559	602	590	499	389	306	206	153	4549
	North MEAN	142	228	365	465	602	638	634	526	394	294	171	104	4560
49	Vancouver	272	285	369	412	474	484	488	445	378	335	276	258	4476
53	Edmonton	255	276	368	419	491	505	507	457	380	328	261	238	4485
50	Regina	268	283	369	413	478	489	492	447	379	333	272	253	4476
50	Winnipeg	268	283	369	413	478	489	492	447	379	333	272	253	4476
46	Sudbury	283	290	370	407	464	471	475	437	376	339	285	270	4467
43	Toronto	292	295	370	402	454	459	465	431	375	342	293	281	4459
45	Montreal	286	292	370	405	461	467	472	435	376	340	287	274	4465
44	Halifax	289	294	370	404	457	463	468	433	375	341	290	278	4462
	South MEAN	277	287	369	409	470	478	482	442	377	336	280	263	4471

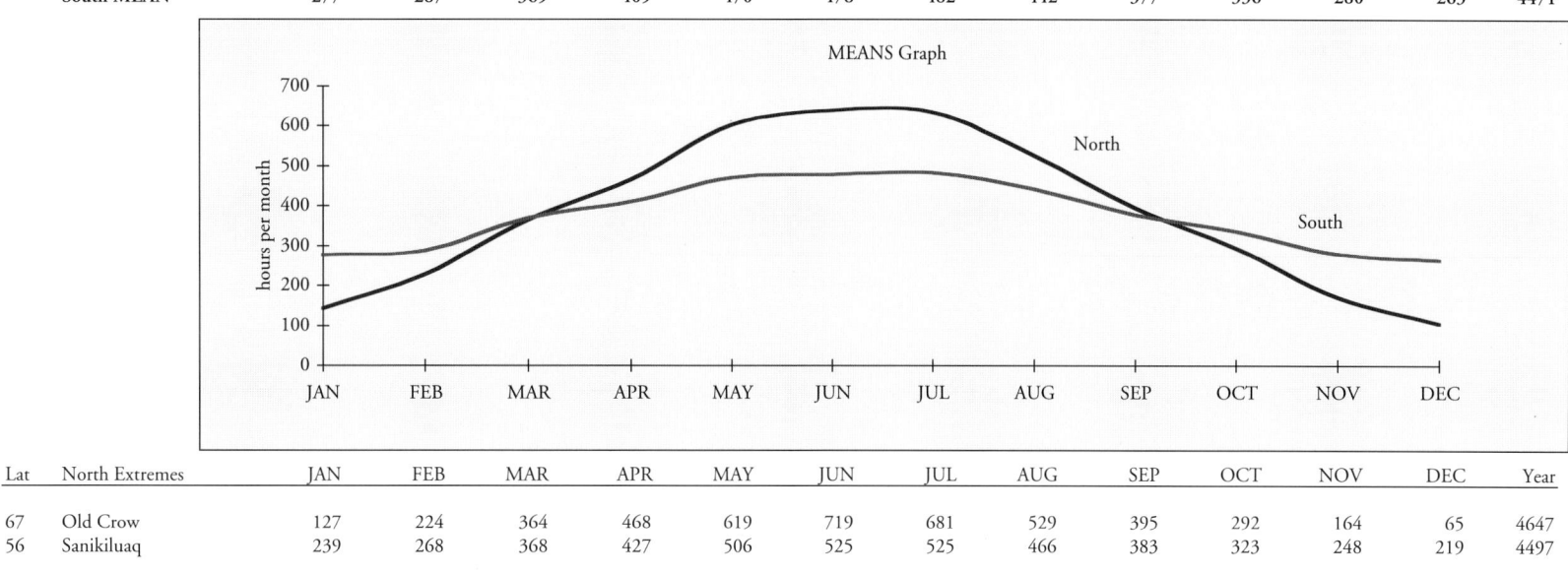

MEANS Graph

Lat	North Extremes	JAN	FEB	MAR	APR	MAY	JUN	JUL	AUG	SEP	OCT	NOV	DEC	Year
67	Old Crow	127	224	364	468	619	719	681	529	395	292	164	65	4647
56	Sanikiluaq	239	268	368	427	506	525	525	466	383	323	248	219	4497

168 APPENDIX A

4.31 TOTAL BRIGHT SUNSHINE Hours per month

1951-80

Lat	Location	JAN	FEB	MAR	APR	MAY	JUN	JUL	AUG	SEP	OCT	NOV	DEC	Year
60	Whitehorse	46	91	153	230	259	273	250	231	137	93	58	23	1844
68	Inuvik	7	65	174	249	295	375	340	216	109	50	18	0	1898
61	Fort Simpson	48	96	160	222	274	281	289	246	134	85	51	29	1915
62	Yellowknife	44	102	195	266	334	395	382	288	152	56	42	21	2277
69	Cambridge Bay	1	52	184	252	258	268	305	176	83	58	10	0	1647
64	Baker Lake	36	107	190	235	264	262	301	211	107	72	51	7	1843
74	Resolute	0	18	146	276	292	256	274	159	59	24	0	0	1504
63	Iqaluit	35	96	177	235	200	175	202	161	82	58	46	20	1487
	North MEAN	**27**	**78**	**172**	**246**	**272**	**286**	**293**	**211**	**108**	**62**	**35**	**13**	**1802**
49	Vancouver	54	87	129	181	246	238	307	256	183	121	69	48	1919
53	Edmonton	98	119	172	233	284	287	313	284	183	163	103	78	2317
50	Regina	100	121	156	209	278	283	342	295	191	168	104	84	2331
50	Winnipeg	121	144	176	220	266	276	316	283	185	151	91	93	2322
46	Sudbury	101	132	152	207	247	246	288	251	151	122	78	85	2060
43	Toronto	92	112	145	182	233	253	281	251	192	149	81	75	2046
45	Montreal	106	128	155	189	242	249	274	240	169	137	86	80	2055
44	Halifax	93	118	140	165	207	203	226	217	182	154	95	85	1885
	South MEAN	**96**	**120**	**153**	**198**	**250**	**254**	**293**	**260**	**180**	**146**	**88**	**79**	**2117**

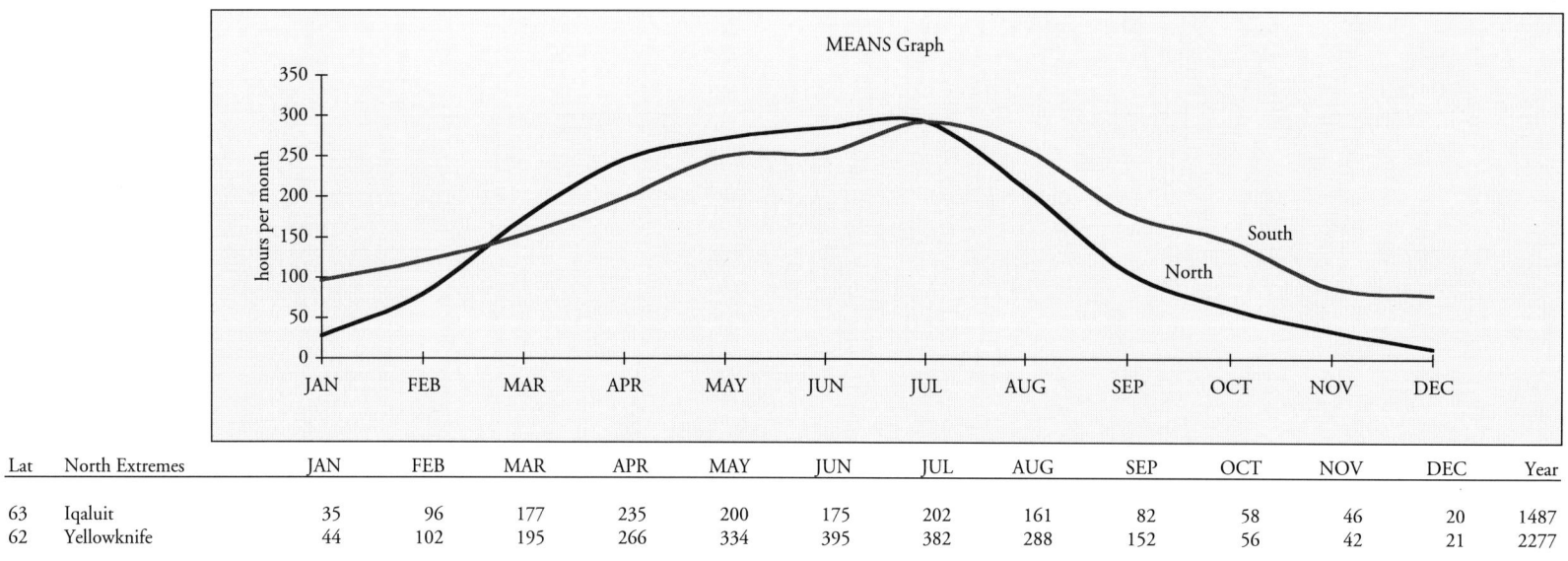

MEANS Graph

Lat	North Extremes	JAN	FEB	MAR	APR	MAY	JUN	JUL	AUG	SEP	OCT	NOV	DEC	Year
63	Iqaluit	35	96	177	235	200	175	202	161	82	58	46	20	1487
62	Yellowknife	44	102	195	266	334	395	382	288	152	56	42	21	2277

APPENDIX A   169

5.31 MEAN DAILY TEMPERATURE   Degrees Celsius

1961-90

Lat	Location	JAN	FEB	MAR	APR	MAY	JUN	JUL	AUG	SEP	OCT	NOV	DEC	Year
60	Whitehorse	-19	-13	-7	0	7	12	14	12	7	1	-10	-16	-1
68	Inuvik	-29	-29	-24	-14	-1	11	14	11	3	-8	-22	-26	-9
61	Fort Simpson	-27	-22	-14	-1	9	15	17	14	8	-2	-17	-24	-4
62	Yellowknife	-28	-25	-19	-6	5	13	17	14	7	-1	-15	-24	-5
69	Cambridge Bay	-33	-34	-31	-22	-10	2	8	6	-1	-12	-24	-30	-15
64	Baker Lake	-33	-32	-28	-18	-7	4	11	9	2	-7	-21	-28	-12
74	Resolute	-32	-33	-31	-24	-11	-1	4	2	-5	-15	-24	-29	-17
63	Iqaluit	-26	-27	-24	-15	-4	3	8	7	2	-5	-13	-22	-10
	North MEAN	**-28**	**-27**	**-22**	**-12**	**-2**	**7**	**12**	**9**	**3**	**-6**	**-18**	**-25**	**-9**
49	Vancouver	3	5	6	9	12	15	17	17	14	10	6	4	10
53	Edmonton	-14	-11	-5	4	10	14	16	15	10	5	-6	-12	2
50	Regina	-17	-13	-6	4	11	16	19	18	12	5	-5	-14	3
50	Winnipeg	-18	-15	-7	4	12	17	20	18	12	6	-5	-15	2
46	Sudbury	-14	-12	-6	3	11	16	19	17	12	6	-1	-10	4
43	Toronto	-7	-6	-1	6	12	17	21	20	15	9	3	-4	7
45	Montreal	-10	-9	-2	6	13	18	21	19	15	8	2	-7	6
44	Halifax	-6	-6	-2	4	9	15	18	18	14	9	3	-3	6
	South MEAN	**-10**	**-8**	**-3**	**5**	**11**	**16**	**19**	**18**	**13**	**7**	**0**	**-8**	**5**

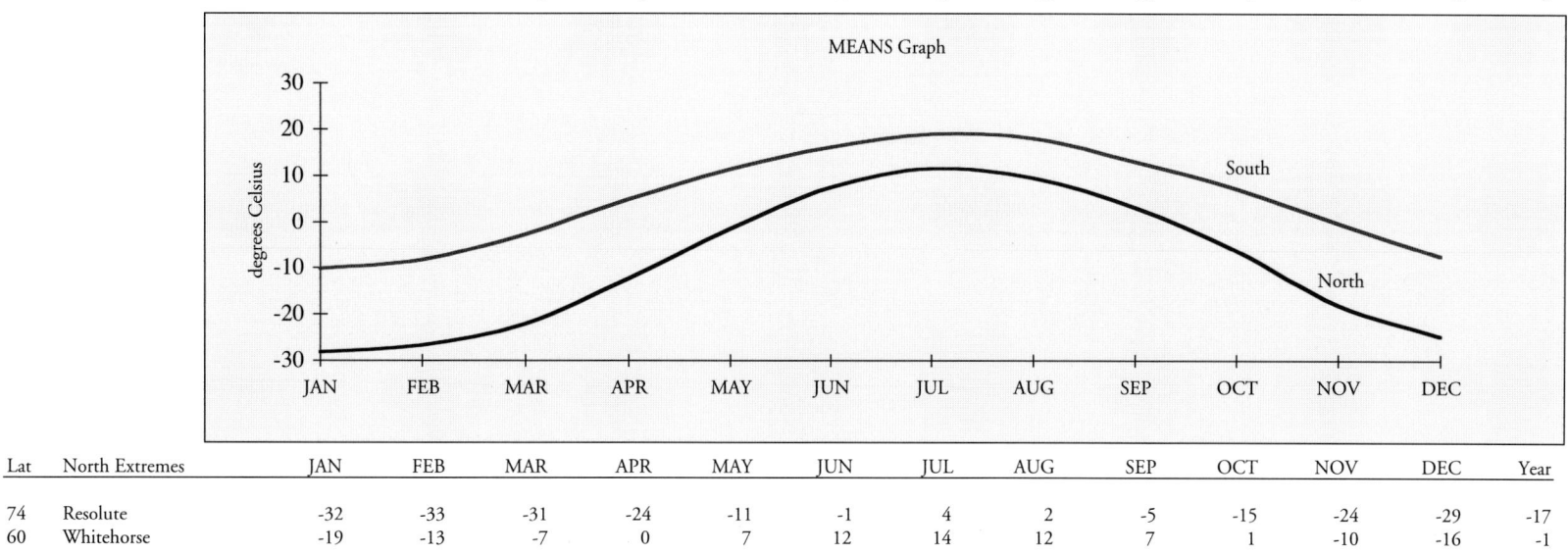

MEANS Graph

Lat	North Extremes	JAN	FEB	MAR	APR	MAY	JUN	JUL	AUG	SEP	OCT	NOV	DEC	Year
74	Resolute	-32	-33	-31	-24	-11	-1	4	2	-5	-15	-24	-29	-17
60	Whitehorse	-19	-13	-7	0	7	12	14	12	7	1	-10	-16	-1

6.31 SOIL TEMPERATURE Degrees Celsius at depth of 1 metre

1951-80

Lat	Location	JAN	FEB	MAR	APR	MAY	JUN	JUL	AUG	SEP	OCT	NOV	DEC	Year
60	Haines Junction	0	-1	-1	-1	0	1	5	7	6	4	2	1	2
68	Inuvik													
61	Fort Simpson	0	-1	-2	-2	-1	0	3	5	5	3	1	0	1
62	Yellowknife													
69	Cambridge Bay													
64	Baker Lake	-15	-18	-18	-16	-10	-2	3	4	2	0	-3	-10	-7
74	Resolute	-16	-18	-19	-19	-17	-11	-3	-2	-2	-6	-10	-13	-11
63	Iqaluit													
	North MEAN	**-8**	**-10**	**-10**	**-10**	**-7**	**-3**	**2**	**4**	**3**	**0**	**-3**	**-6**	**-4**
49	Vancouver	7	7	7	9	11	14	15	16	16	14	11	9	11
53	Edmonton													
50	Regina	0	-1	-1	-1	1	6	10	13	12	9	6	2	5
50	Winnipeg	1	0	0	0	2	8	12	14	13	11	7	3	6
46	Sudbury													
43	Toronto	4	2	2	3	8	12	15	17	16	13	9	6	9
45	Montreal	2	1	1	1	6	10	14	15	14	11	7	4	7
44	Halifax													
	South MEAN	**3**	**2**	**2**	**2**	**6**	**10**	**13**	**15**	**14**	**12**	**8**	**5**	**8**

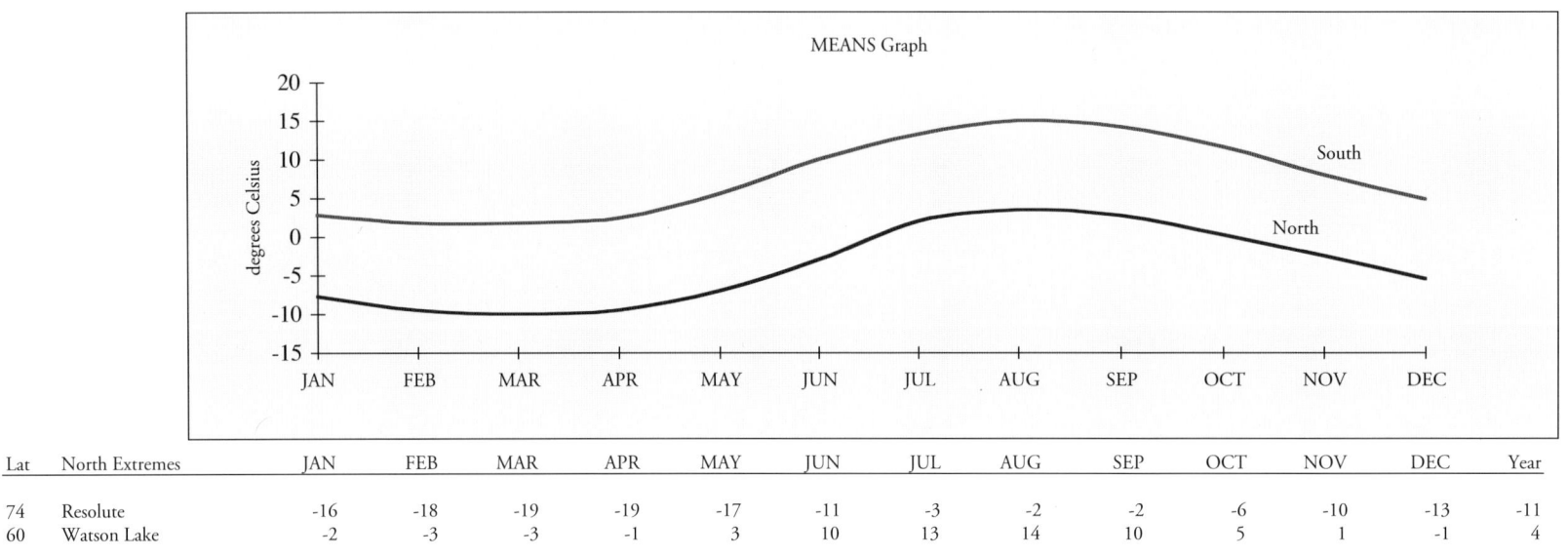

MEANS Graph

Lat	North Extremes	JAN	FEB	MAR	APR	MAY	JUN	JUL	AUG	SEP	OCT	NOV	DEC	Year
74	Resolute	-16	-18	-19	-19	-17	-11	-3	-2	-2	-6	-10	-13	-11
60	Watson Lake	-2	-3	-3	-1	3	10	13	14	10	5	1	-1	4

7.31 NUMBER OF DAYS WITH FROST

1951-80

Lat	Location	JAN	FEB	MAR	APR	MAY	JUN	JUL	AUG	SEP	OCT	NOV	DEC	Year
60	Whitehorse	31	28	30	28	15	2	0	1	8	22	28	31	224
68	Inuvik	31	28	31	30	27	6	1	3	18	31	30	31	267
61	Fort Simpson	31	28	31	27	11	1	0	1	12	28	30	31	231
62	Yellowknife	31	28	31	28	14	1	0	0	7	25	30	31	226
69	Cambridge Bay	31	28	31	30	31	20	1	4	24	31	30	31	292
64	Baker Lake	31	28	31	30	30	14	0	1	19	30	30	31	275
74	Resolute	31	28	31	30	31	24	10	17	29	31	30	31	323
63	Iqaluit	31	28	31	30	29	14	1	1	17	30	30	31	273
	North MEAN	**31**	**28**	**31**	**29**	**24**	**10**	**2**	**4**	**17**	**29**	**30**	**31**	**264**
49	Vancouver	15	10	9	1	0	0	0	0	0	1	8	11	55
53	Edmonton	31	28	30	24	7	1	0	0	6	22	30	31	210
50	Regina	31	28	30	22	7	1	0	0	5	20	29	31	204
50	Winnipeg	31	28	30	21	8	0	0	0	3	15	28	31	195
46	Sudbury	31	28	30	21	6	0	0	0	2	11	24	30	183
43	Toronto	30	27	27	14	3	0	0	0	1	7	18	28	155
45	Montreal	31	28	28	13	1	0	0	0	0	7	18	29	155
44	Halifax	30	27	29	21	4	0	0	0	0	5	19	28	163
	South MEAN	**29**	**26**	**27**	**17**	**5**	**0**	**0**	**0**	**2**	**11**	**22**	**27**	**165**

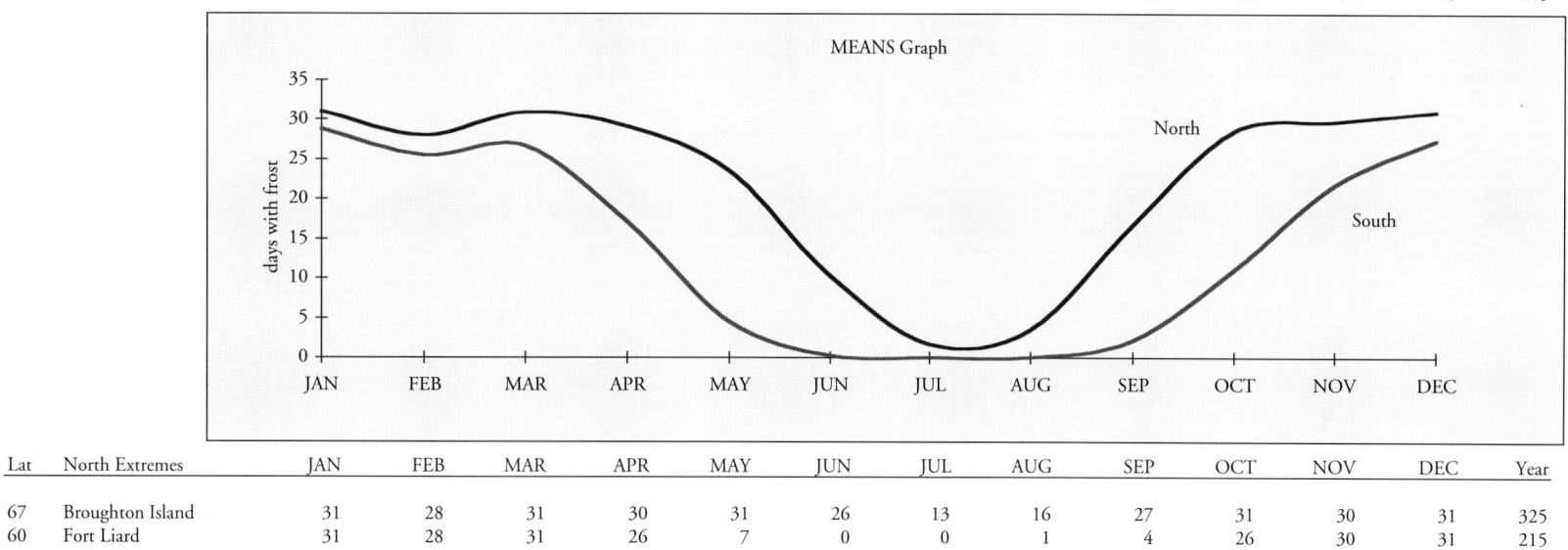

MEANS Graph

Lat	North Extremes	JAN	FEB	MAR	APR	MAY	JUN	JUL	AUG	SEP	OCT	NOV	DEC	Year
67	Broughton Island	31	28	31	30	31	26	13	16	27	31	30	31	325
60	Fort Liard	31	28	31	26	7	0	0	1	4	26	30	31	215

8.31 DEGREE-DAYS BELOW 18° C

1961-90

Lat	Location	JAN	FEB	MAR	APR	MAY	JUN	JUL	AUG	SEP	OCT	NOV	DEC	Year
60	Whitehorse	1142	880	784	532	354	193	128	178	322	538	842	1054	6947
68	Inuvik	1453	1315	1310	965	581	226	142	234	441	817	1187	1371	10040
61	Fort Simpson	1387	1134	1000	582	296	108	59	128	314	623	1046	1300	7976
62	Yellowknife	1428	1202	1134	728	404	151	67	131	338	603	986	1308	8477
69	Cambridge Bay	1597	1456	1512	1202	854	485	311	366	559	919	1253	1478	11991
64	Baker Lake	1573	1415	1425	1077	766	417	216	266	469	791	1161	1435	11011
74	Resolute	1555	1442	1529	1249	903	560	433	500	693	1033	1272	1461	12630
63	Iqaluit	1360	1267	1288	984	689	439	320	347	473	713	923	1246	10050
	North MEAN	**1437**	**1264**	**1248**	**915**	**606**	**322**	**209**	**269**	**451**	**755**	**1084**	**1332**	**9890**
49	Vancouver	466	376	363	276	183	89	40	36	113	249	360	451	3002
53	Edmonton	1001	816	728	431	239	120	75	105	244	417	713	940	5827
50	Regina	1073	874	748	418	212	76	26	50	200	400	695	984	5756
50	Winnipeg	1129	936	779	429	212	69	19	46	179	381	683	1014	5874
46	Sudbury	980	847	733	451	230	88	26	55	179	371	580	868	5407
43	Toronto	770	682	583	360	187	57	12	23	106	283	445	667	4174
45	Montreal	880	759	634	370	167	44	9	25	120	301	492	775	4575
44	Halifax	742	679	612	433	267	107	28	33	130	294	444	652	4422
	South MEAN	**880**	**746**	**647**	**396**	**212**	**81**	**29**	**47**	**159**	**337**	**551**	**794**	**4880**

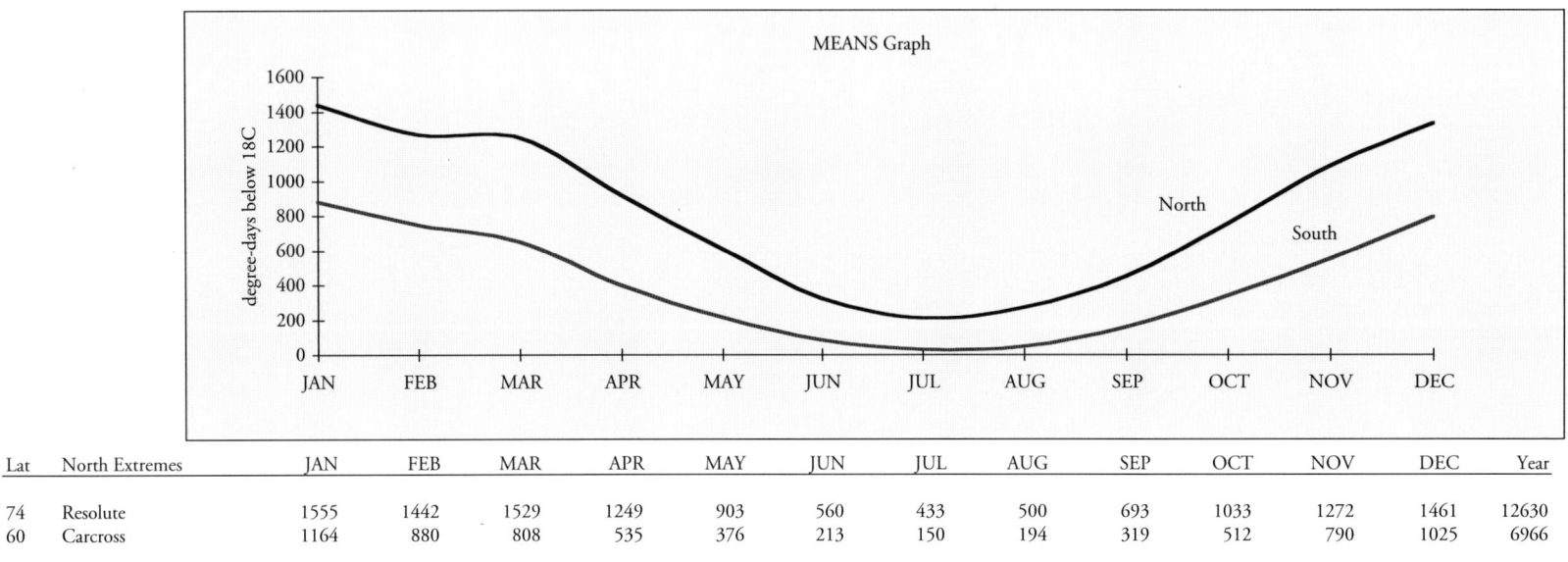

MEANS Graph

Lat	North Extremes	JAN	FEB	MAR	APR	MAY	JUN	JUL	AUG	SEP	OCT	NOV	DEC	Year
74	Resolute	1555	1442	1529	1249	903	560	433	500	693	1033	1272	1461	12630
60	Carcross	1164	880	808	535	376	213	150	194	319	512	790	1025	6966

9.31 MEAN WIND SPEED Kilometres per hour

1961-90

Lat	Location	JAN	FEB	MAR	APR	MAY	JUN	JUL	AUG	SEP	OCT	NOV	DEC	Year
60	Whitehorse	13	15	14	14	14	13	11	12	14	16	15	14	14
68	Inuvik	7	7	9	10	12	13	12	11	11	10	7	7	10
61	Fort Simpson	8	9	10	11	11	10	9	9	9	10	9	7	9
62	Yellowknife	13	13	14	16	16	16	15	15	15	16	15	12	15
69	Cambridge Bay	23	22	21	20	21	20	20	22	23	23	21	21	21
64	Baker Lake	24	23	22	21	20	18	18	18	20	22	23	22	21
74	Resolute	21	21	20	20	21	21	20	21	24	23	22	21	21
63	Iqaluit	16	16	14	17	18	16	13	14	16	18	18	15	16
	North MEAN	**16**	**16**	**16**	**16**	**17**	**16**	**15**	**15**	**17**	**17**	**16**	**15**	**16**
49	Vancouver	12	12	13	13	11	11	11	11	10	11	12	12	12
53	Edmonton	13	13	13	15	16	14	11	11	13	13	12	13	13
50	Regina	21	21	21	22	21	19	17	17	19	20	19	20	20
50	Winnipeg	18	17	18	20	19	17	15	15	18	19	18	17	18
46	Sudbury	19	19	20	20	18	18	16	15	17	18	19	19	18
43	Toronto	18	17	17	17	15	13	12	11	12	13	16	16	15
45	Montreal	17	16	16	16	14	14	12	11	12	14	16	16	15
44	Halifax	19	19	20	19	18	17	16	15	15	17	18	19	18
	South MEAN	**17**	**17**	**17**	**18**	**17**	**15**	**14**	**13**	**15**	**16**	**16**	**17**	**16**

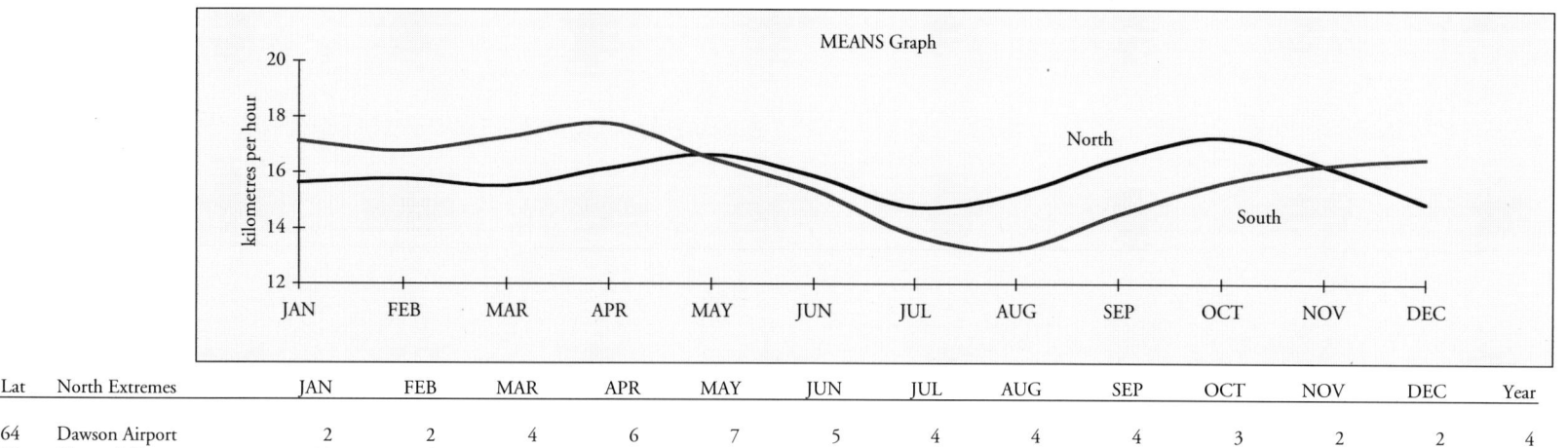

MEANS Graph

Lat	North Extremes	JAN	FEB	MAR	APR	MAY	JUN	JUL	AUG	SEP	OCT	NOV	DEC	Year
64	Dawson Airport	2	2	4	6	7	5	4	4	4	3	2	2	4
63	Chesterfield	23	25	23	21	21	19	17	20	23	28	24	24	22

MAXIMUM GUST SPEED   150 kph or more
- 74 Resolute   158 E
- 64 Baker Lake   177 WNW
- 63 Iqaluit   156 NW
- 61 Burwash   171 S
- 60 Watson Lake   150 SW

APPENDIX A

10.31 PREVAILING WIND DIRECTION By month; percent calm

1951-80

Lat	Location	JAN	FEB	MAR	APR	MAY	JUN	JUL	AUG	SEP	OCT	NOV	DEC	Calm
60	Whitehorse	SSE	SSE	SSE	SE	SE	SE	SE	SE	SSE	SSE	S	SSE	12
68	Inuvik	E	E	E	E	ENE	E	E	E	E	E	E	E	14
61	Fort Simpson	NNW	NNW	NNW	NNWE	NNW	NNW	NNW	NNW	SE	SE	SE	NW	19
62	Yellowknife	NW	E	E	E	E	S	S	S	E	E	E	E	13
69	Cambridge Bay	NW	W	NW	NE	NW	N	W	W	N	NW	NW	W	13
64	Baker Lake	N	N	N	N	N	N	N	N	NNW	N	N	N	17
74	Resolute	NNW	NNW	NNW	NNW	NNW	NNW	W	SE	N	NNW	NNW	NNW	13
63	Iqaluit	NW	NW	NW	NW	NW	NW	SSE	SE	NW	NW	NW	NW	23
49	Vancouver	E	E	E	E	E	E	E	E	E	E	E	E	23
53	Edmonton	S	S	S	S	SE	W	W	W	S	S	S	S	12
50	Regina	SE	SE	SE	SE	SE	SE	SE	SE	SE	SE	SE	SE	14
50	Winnipeg	S	S	S	S	S	S	S	S	S	S	S	S	15
46	Sudbury	N	N	N	N	N	SW	SW	SW	S	S	S	N	10
43	Toronto	WSW	N	N	N	N	N	N	N	N	W	W	W	11
45	Montreal	WSW	WSW	WSW	W	SW	SW	SW	SW	SW	W	W	W	14
44	Halifax	WNW	WNW	N	N	S	S	SSW	S	SSW	SSW	N	NW	3

11.31 MEAN WIND CHILL COOLING RATE Watts per square metre

1951-80

Lat	Location	JAN	FEB	MAR	APR	MAY	JUN	JUL	AUG	SEP	OCT	NOV	DEC	Year
60	Whitehorse	1600	1400	1200	1000	800	600	500	500	700	1000	1300	1600	1017
68	Inuvik	1700	1600	1600	1400	900	700	500	600	900	1100	1400	1600	1167
61	Fort Simpson	1600	1500	1400	1000	700	500	400	500	700	1000	1300	1500	1008
62	Yellowknife	1900	1800	1600	1300	900	600	500	600	800	1100	1400	1700	1183
69	Cambridge Bay	2300	2200	2100	1800	1400	1000	800	900	1200	1500	1900	2100	1600
64	Baker Lake	2300	2300	2100	1700	1300	1000	700	700	1000	1400	1800	2000	1525
74	Resolute	2200	2200	2100	1800	1500	1200	1000	1000	1300	1600	1900	2100	1658
63	Iqaluit	1800	1800	1700	1500	1200	1000	800	800	1000	1300	1500	1700	1342
	North MEAN	**1925**	**1850**	**1725**	**1438**	**1088**	**825**	**650**	**700**	**950**	**1250**	**1563**	**1788**	**1313**
49	Vancouver	900	800	800	700	600	500	400	400	500	600	800	900	658
53	Edmonton	1500	1300	1200	900	700	500	400	500	600	800	1100	1300	900
50	Regina	1700	1600	1400	1000	700	500	400	400	700	900	1300	1500	1008
50	Winnipeg	1700	1600	1300	1000	700	500	400	400	600	900	1300	1500	992
46	Sudbury	1600	1500	1300	1000	700	500	400	500	700	900	1100	1400	967
43	Toronto	1300	1300	1100	800	600	400	300	300	500	700	1000	1200	792
45	Montreal	1400	1300	1200	800	600	400	300	300	500	700	1000	1300	817
44	Halifax	1300	1300	1200	1000	800	500	400	400	600	800	1000	1200	875
	South MEAN	**1425**	**1338**	**1188**	**900**	**675**	**475**	**375**	**400**	**588**	**788**	**1075**	**1288**	**876**

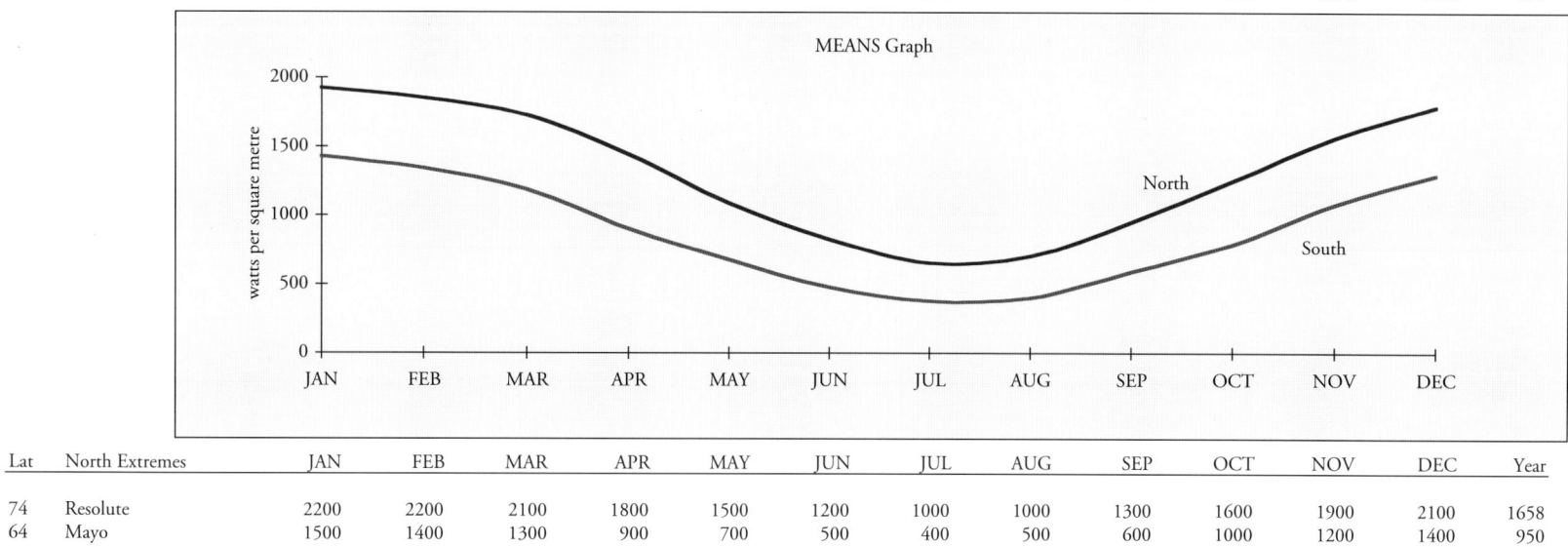

MEANS Graph

Lat	North Extremes	JAN	FEB	MAR	APR	MAY	JUN	JUL	AUG	SEP	OCT	NOV	DEC	Year
74	Resolute	2200	2200	2100	1800	1500	1200	1000	1000	1300	1600	1900	2100	1658
64	Mayo	1500	1400	1300	900	700	500	400	500	600	1000	1200	1400	950

12.31 TOTAL PRECIPITATION Millimetres

1961-90

Lat	Location	JAN	FEB	MAR	APR	MAY	JUN	JUL	AUG	SEP	OCT	NOV	DEC	Year
60	Whitehorse	17	12	12	8	14	31	39	39	35	23	19	19	269
68	Inuvik	16	11	11	13	19	22	34	44	24	30	18	17	258
61	Fort Simpson	20	18	18	16	30	44	53	51	30	36	26	19	360
62	Yellowknife	15	13	11	10	17	23	35	42	29	35	24	15	267
69	Cambridge Bay	4	4	5	7	11	12	22	29	20	15	7	5	141
64	Baker Lake	8	7	11	16	14	22	40	40	41	35	19	9	262
74	Resolute	4	3	5	6	8	13	23	32	23	13	6	5	140
63	Iqaluit	22	19	22	28	30	37	58	64	52	42	31	20	424
	North MEAN	**13**	**11**	**12**	**13**	**18**	**26**	**38**	**42**	**32**	**29**	**19**	**13**	**265**
49	Vancouver	150	124	109	75	62	46	36	38	64	115	170	179	1167
53	Edmonton	23	16	16	22	43	76	101	70	48	18	16	19	466
50	Regina	15	13	17	20	51	67	59	40	34	20	12	16	364
50	Winnipeg	19	15	23	36	60	84	72	75	51	30	21	19	505
46	Sudbury	60	49	61	63	71	84	71	87	103	76	79	66	872
43	Toronto	46	46	57	64	66	69	77	84	74	63	70	66	781
45	Montreal	63	56	68	75	68	83	86	100	87	75	93	86	940
44	Halifax	147	119	123	124	111	98	97	110	95	129	154	167	1474
	South MEAN	**65**	**55**	**59**	**60**	**66**	**76**	**75**	**76**	**70**	**66**	**77**	**77**	**821**

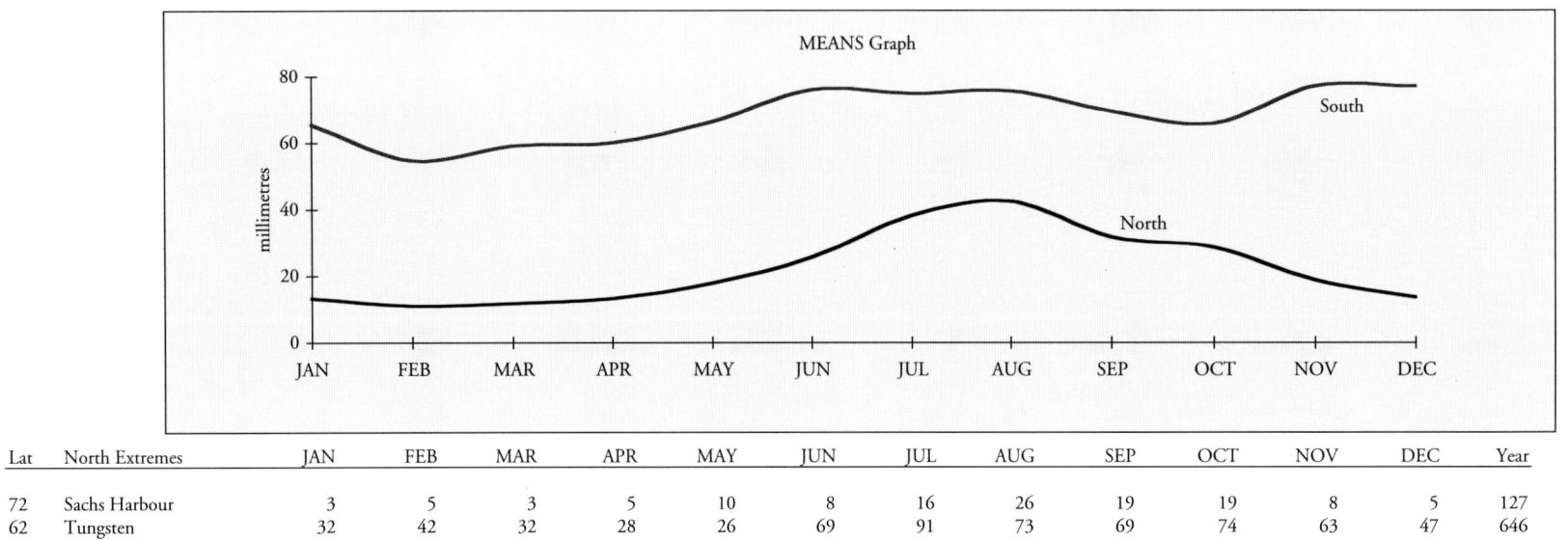

MEANS Graph

Lat	North Extremes	JAN	FEB	MAR	APR	MAY	JUN	JUL	AUG	SEP	OCT	NOV	DEC	Year
72	Sachs Harbour	3	5	3	5	10	8	16	26	19	19	8	5	127
62	Tungsten	32	42	32	28	26	69	91	73	69	74	63	47	646

13.31 MEAN SNOWFALL centimetres

1961-90

Lat	Location	JAN	FEB	MAR	APR	MAY	JUN	JUL	AUG	SEP	OCT	NOV	DEC	Year
60	Whitehorse	23.0	16.6	16.9	9.8	2.9	0.9	0.0	1.1	4.8	18.7	25.5	25.1	145.3
68	Inuvik	19.1	14.1	14.2	15.6	15.0	2.3	0.4	4.0	10.8	34.8	23.7	21.3	175.3
61	Fort Simpson	21.7	19.7	19.4	14.2	7.7	0.0	0.0	0.6	4.6	26.5	28.2	21.5	164.1
62	Yellowknife	18.8	17.1	13.7	10.5	4.3	0.2	0.0	0.0	3.5	21.7	33.5	20.6	143.9
69	Cambridge Bay	4.9	5.3	6.0	8.3	10.6	3.4	0.0	1.0	9.2	15.8	9.1	6.0	79.6
64	Baker Lake	9.1	7.7	12.3	16.7	8.7	4.0	0.0	1.3	7.7	29.3	22.1	11.2	130.1
74	Resolute	3.8	3.5	5.0	6.8	10.0	8.1	4.7	10.3	18.6	14.6	6.5	5.3	97.2
63	Iqaluit	24.4	21.4	25.3	32.5	28.2	11.7	0.3	0.7	14.4	38.8	35.2	23.9	256.8
	North MEAN	**15.6**	**13.2**	**14.1**	**14.3**	**10.9**	**3.8**	**0.7**	**2.4**	**9.2**	**25.0**	**23.0**	**16.9**	**149.0**
49	Vancouver	20.6	8.6	4.1	0.5	0.0	0.0	0.0	0.0	0.0	0.0	2.6	18.6	55.0
53	Edmonton	25.7	18.5	17.7	12.2	3.6	0.0	0.0	0.0	2.5	7.8	16.7	22.4	127.1
50	Regina	19.2	16.3	17.0	8.8	3.0	0.0	0.0	0.0	1.8	7.6	12.6	21.2	107.5
50	Winnipeg	22.6	17.1	19.2	9.4	2.0	0.0	0.0	0.0	0.4	4.9	19.0	20.1	114.7
46	Sudbury	59.5	48.9	37.9	18.1	1.7	0.0	0.0	0.0	0.1	6.7	32.6	61.2	266.7
43	Toronto	32.3	25.9	19.9	7.3	0.1	0.0	0.0	0.0	0.0	1.1	6.4	31.1	124.1
45	Montreal	47.7	41.2	31.3	10.9	1.6	0.0	0.0	0.0	0.0	2.6	24.1	54.8	214.2
44	Halifax	64.2	59.9	44.4	22.1	3.1	0.0	0.0	0.0	0.0	2.7	14.4	50.6	261.4
	South MEAN	**36.5**	**29.6**	**23.9**	**11.2**	**1.9**	**0.0**	**0.0**	**0.0**	**0.6**	**4.2**	**16.1**	**35.0**	**158.8**

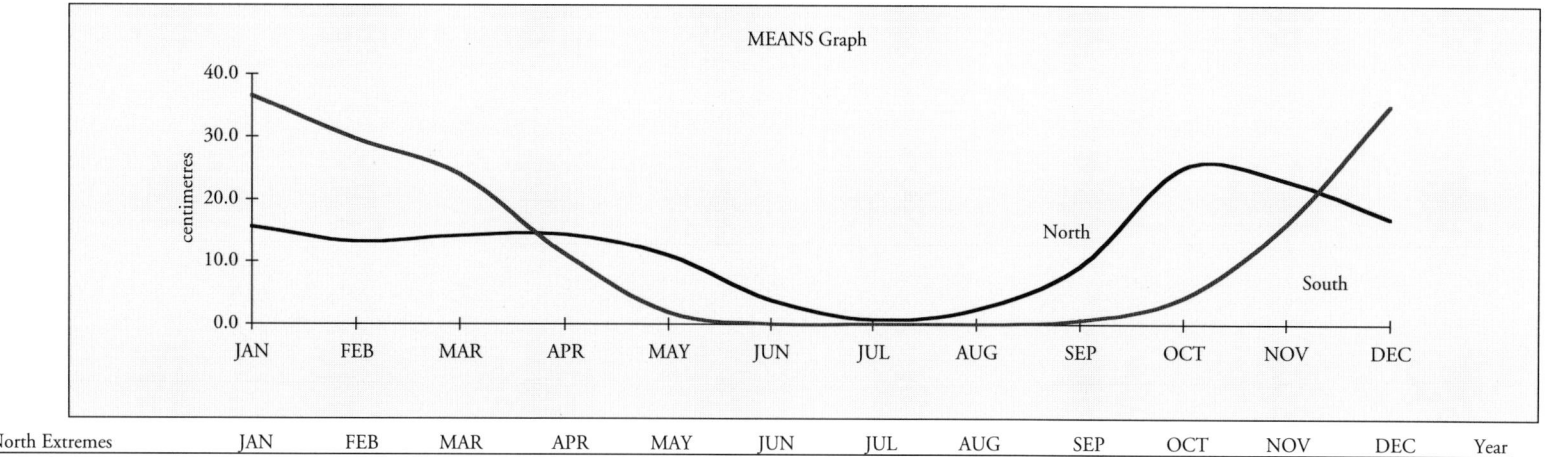

Lat	North Extremes	JAN	FEB	MAR	APR	MAY	JUN	JUL	AUG	SEP	OCT	NOV	DEC	Year
73	Arctic Bay	4.4	3.5	5.0	6.9	7.9	3.3	0.3	1.9	12.0	16.9	5.4	4.0	71.5
67	Broughton Island	9.0	9.3	5.5	15.4	31.9	20.5	7.6	9.8	34.1	52.5	37.1	10.7	243.4
64	Keno Hill	36.2	49.6	36.0	26.7	21.4	2.6	0.0	0.2	45.6	57.2	41.0	49.2	365.7

APPENDIX A

14.31 NUMBER OF DAYS WITH BLOWING SNOW

1951-80

Lat	Location	JAN	FEB	MAR	APR	MAY	JUN	JUL	AUG	SEP	OCT	NOV	DEC	Year
60	Whitehorse	1	0	0	0	0	0	0	0	0	0	1	1	3
68	Inuvik	2	1	1	1	0	0	0	0	0	1	1	2	9
61	Fort Simpson	2	1	1	1	0	0	0	0	0	1	1	1	8
62	Yellowknife	3	1	1	1	0	0	0	0	0	1	1	2	10
69	Cambridge Bay	14	10	9	6	4	1	0	0	1	7	8	10	70
64	Baker Lake	17	14	13	10	5	0	0	0	1	6	12	13	91
74	Resolute	14	12	11	8	6	1	0	0	4	11	11	13	91
63	Iqaluit	11	10	7	7	2	0	0	0	1	5	8	10	61
	North MEAN	**8**	**6**	**5**	**4**	**2**	**0**	**0**	**0**	**1**	**4**	**5**	**7**	**43**
49	Vancouver	0	0	0	0	0	0	0	0	0	0	0	0	0
53	Edmonton	3	2	2	0	0	0	0	0	0	0	0	2	9
50	Regina	10	8	5	1	0	0	0	0	0	0	2	6	32
50	Winnipeg	7	6	4	1	0	0	0	0	0	0	2	5	25
46	Sudbury	5	4	3	1	0	0	0	0	0	0	2	4	19
43	Toronto	3	2	1	0	0	0	0	0	0	0	0	2	8
45	Montreal	5	5	2	0	0	0	0	0	0	0	1	3	16
44	Halifax	3	4	3	1	0	0	0	0	0	0	0	3	14
	South MEAN	**5**	**4**	**3**	**1**	**0**	**0**	**0**	**0**	**0**	**0**	**1**	**3**	**15**

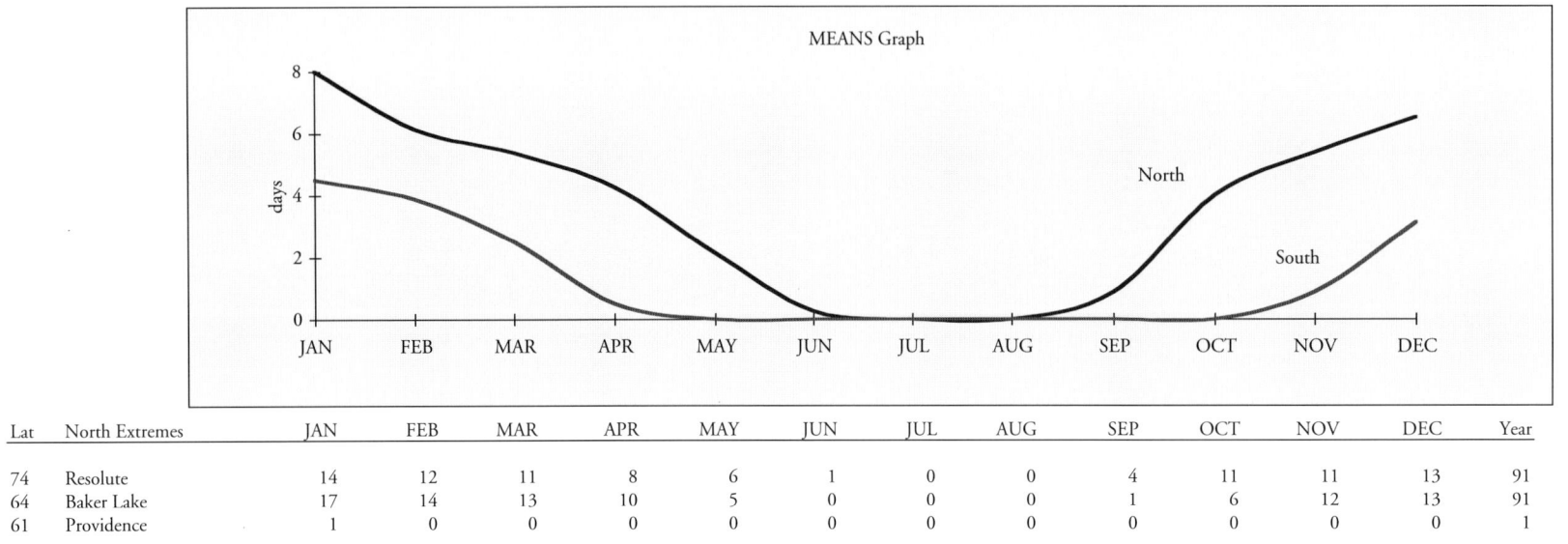

MEANS Graph

Lat	North Extremes	JAN	FEB	MAR	APR	MAY	JUN	JUL	AUG	SEP	OCT	NOV	DEC	Year
74	Resolute	14	12	11	8	6	1	0	0	4	11	11	13	91
64	Baker Lake	17	14	13	10	5	0	0	0	1	6	12	13	91
61	Providence	1	0	0	0	0	0	0	0	0	0	0	0	1

15.31 RELATIVE HUMIDITY Percent

1951-80

Lat	Location	JAN	FEB	MAR	APR	MAY	JUN	JUL	AUG	SEP	OCT	NOV	DEC	Year
60	Whitehorse	77	73	68	60	53	54	60	65	71	73	79	79	68
68	Inuvik	71	70	68	71	72	63	65	73	79	84	78	74	72
61	Fort Simpson	77	78	74	67	62	62	66	71	78	85	83	79	74
62	Yellowknife	73	73	70	70	61	57	60	69	76	84	82	75	71
69	Cambridge Bay	75	78	75	79	86	86	79	85	88	87	81	78	81
64	Baker Lake	65	66	66	74	82	77	72	76	82	83	75	69	74
74	Resolute	68	68	67	71	81	86	85	88	88	82	72	69	77
63	Iqaluit	73	73	73	76	79	76	76	78	79	81	78	74	76
	North MEAN	**72**	**72**	**70**	**71**	**72**	**70**	**70**	**76**	**80**	**82**	**79**	**75**	**74**
49	Vancouver	85	85	81	77	74	75	74	79	83	87	86	87	81
53	Edmonton	73	75	75	67	58	65	73	75	75	70	76	76	72
50	Regina	79	81	81	70	60	61	63	63	67	69	79	81	71
50	Winnipeg	79	81	80	70	61	65	68	69	72	73	81	81	73
46	Sudbury	77	76	73	69	64	66	69	73	78	79	83	81	74
43	Toronto	80	79	78	71	68	70	70	74	77	78	81	82	76
45	Montreal	75	74	72	67	64	69	70	73	76	75	77	78	73
44	Halifax	83	80	79	78	75	77	81	80	81	82	84	84	80
	South MEAN	**79**	**79**	**77**	**71**	**66**	**69**	**71**	**73**	**76**	**77**	**81**	**81**	**75**

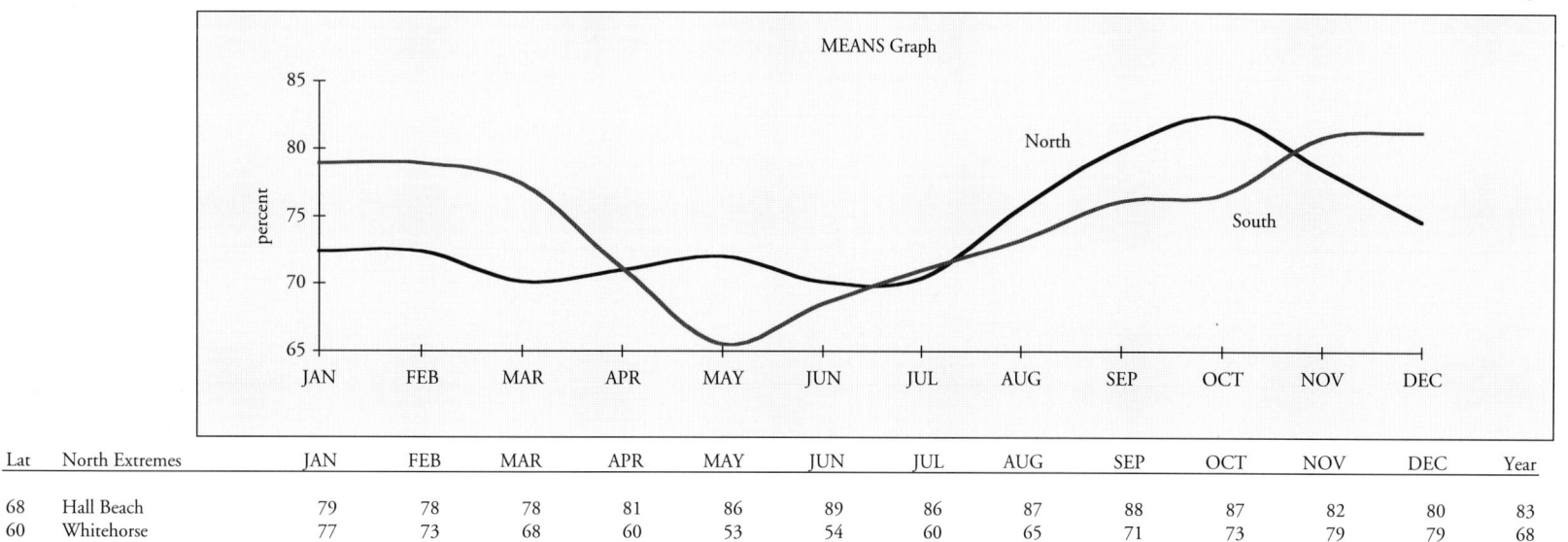

MEANS Graph

Lat	North Extremes	JAN	FEB	MAR	APR	MAY	JUN	JUL	AUG	SEP	OCT	NOV	DEC	Year
68	Hall Beach	79	78	78	81	86	89	86	87	88	87	82	80	83
60	Whitehorse	77	73	68	60	53	54	60	65	71	73	79	79	68

16.31 VAPOUR PRESSURE KiloPascals

1951-80

Lat	Location	JAN	FEB	MAR	APR	MAY	JUN	JUL	AUG	SEP	OCT	NOV	DEC	Year
60	Whitehorse	0.16	0.22	0.26	0.37	0.51	0.74	0.92	0.90	0.70	0.48	0.29	0.20	0.48
68	Inuvik	0.08	0.07	0.08	0.19	0.44	0.76	1.00	0.93	0.61	0.32	0.13	0.08	0.39
61	Fort Simpson	0.08	0.12	0.18	0.39	0.68	1.03	1.24	1.13	0.78	0.47	0.18	0.10	0.53
62	Yellowknife	0.07	0.09	0.13	0.30	0.54	0.84	1.10	1.09	0.76	0.49	0.22	0.10	0.48
69	Cambridge Bay	0.05	0.05	0.05	0.11	0.30	0.60	0.84	0.82	0.53	0.26	0.09	0.06	0.31
64	Baker Lake	0.04	0.05	0.05	0.15	0.35	0.64	0.94	0.93	0.61	0.34	0.13	0.06	0.36
74	Resolute	0.04	0.04	0.04	0.08	0.24	0.51	0.68	0.64	0.40	0.19	0.08	0.05	0.25
63	Iqaluit	0.07	0.09	0.13	0.30	0.54	0.84	1.10	1.09	0.76	0.49	0.22	0.10	0.48
	North MEAN	**0.07**	**0.09**	**0.12**	**0.24**	**0.45**	**0.75**	**0.98**	**0.94**	**0.64**	**0.38**	**0.17**	**0.09**	**0.41**
49	Vancouver	0.66	0.73	0.74	0.86	1.05	1.28	1.46	1.50	1.33	1.06	0.82	0.73	1.02
53	Edmonton	0.18	0.24	0.32	0.50	0.71	1.05	1.31	1.25	0.87	0.57	0.34	0.22	0.63
50	Regina	0.17	0.22	0.33	0.53	0.78	1.14	1.37	1.23	0.87	0.59	0.36	0.23	0.65
50	Winnipeg	0.14	0.19	0.31	0.55	0.81	1.25	1.57	1.44	1.02	0.69	0.39	0.22	0.72
46	Sudbury	0.21	0.23	0.32	0.53	0.81	1.21	1.46	1.43	1.12	0.79	0.51	0.28	0.74
43	Toronto	0.34	0.35	0.47	0.67	1.00	1.45	1.67	1.66	1.35	0.94	0.67	0.43	0.92
45	Montreal	0.26	0.28	0.40	0.62	0.98	1.46	1.73	1.66	1.30	0.87	0.59	0.33	0.87
44	Halifax	0.38	0.35	0.45	0.59	0.85	1.25	1.62	1.60	1.28	0.95	0.70	0.46	0.87
	South MEAN	**0.29**	**0.32**	**0.42**	**0.61**	**0.87**	**1.26**	**1.52**	**1.47**	**1.14**	**0.81**	**0.55**	**0.36**	**0.80**

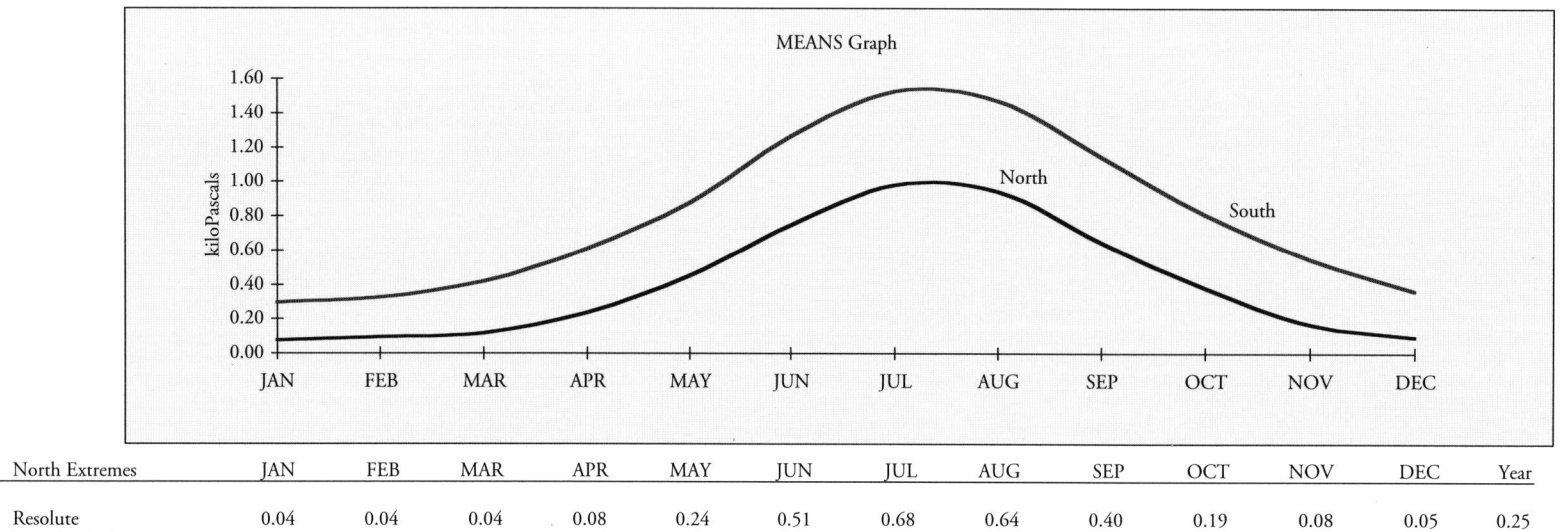

MEANS Graph

Lat	North Extremes	JAN	FEB	MAR	APR	MAY	JUN	JUL	AUG	SEP	OCT	NOV	DEC	Year
74	Resolute	0.04	0.04	0.04	0.08	0.24	0.51	0.68	0.64	0.40	0.19	0.08	0.05	0.25
61	Fort Resolution	0.07	0.11	0.17	0.35	0.62	0.99	1.31	1.26	0.84	0.56	0.20	0.12	0.55

17.31 NUMBER OF DAYS WITH FOG

1961-90

Lat	Location	JAN	FEB	MAR	APR	MAY	JUN	JUL	AUG	SEP	OCT	NOV	DEC	Year
60	Whitehorse	4	1	0	0	0	0	0	1	2	2	2	3	15
68	Inuvik	2	1	1	1	4	2	1	3	4	3	1	1	24
61	Fort Simpson	0	0	0	0	1	1	1	3	4	6	3	1	20
62	Yellowknife	2	2	1	0	1	1	0	1	2	4	3	2	19
69	Cambridge Bay	4	5	4	4	7	5	4	4	5	6	3	3	54
64	Baker Lake	6	6	4	3	4	2	2	1	2	4	3	4	41
74	Resolute	3	4	5	3	4	9	12	13	7	5	2	2	69
63	Iqaluit	1	1	0	0	1	1	3	3	2	1	1	1	15
	North MEAN	**3**	**3**	**2**	**1**	**3**	**3**	**3**	**4**	**4**	**4**	**2**	**2**	**32**
49	Vancouver	5	4	1	0	0	0	1	1	5	7	5	5	34
53	Edmonton	1	2	2	1	1	1	1	2	1	1	3	2	18
50	Regina	3	4	4	2	1	1	1	1	2	2	3	4	28
50	Winnipeg	2	2	3	1	1	1	0	0	1	2	2	2	17
46	Sudbury	4	4	5	5	5	5	4	7	8	8	8	5	68
43	Toronto	2	3	3	2	3	2	2	3	3	4	3	4	34
45	Montreal	1	2	1	1	1	1	1	1	2	2	3	2	18
44	Halifax	6	6	8	10	13	15	17	14	9	8	8	7	121
	South MEAN	**3**	**3**	**3**	**3**	**3**	**3**	**3**	**4**	**4**	**4**	**4**	**4**	**42**

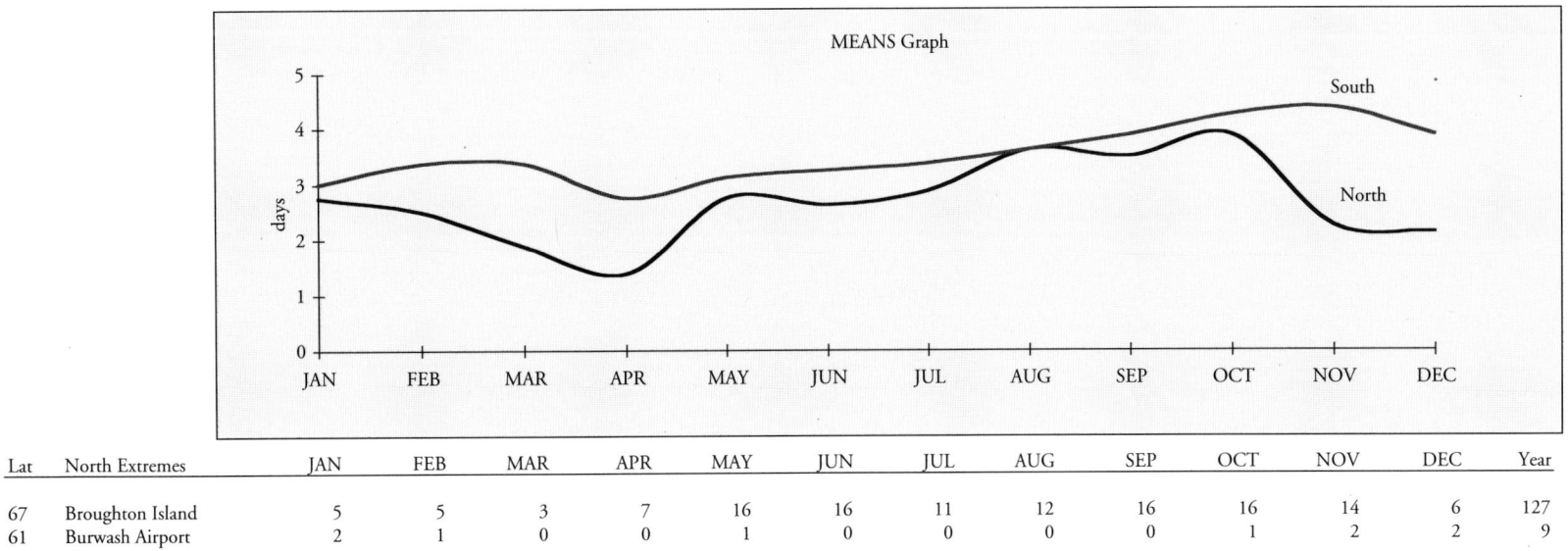

MEANS Graph

Lat	North Extremes	JAN	FEB	MAR	APR	MAY	JUN	JUL	AUG	SEP	OCT	NOV	DEC	Year
67	Broughton Island	5	5	3	7	16	16	11	12	16	16	14	6	127
61	Burwash Airport	2	1	0	0	1	0	0	0	0	1	2	2	9

18.31 DESIGN DATA from the National Building Code of Canada 1995

Lat. °N	Location	Elev. m	Design Temperatures January 2.5% °C	1% °C	July 2.5% Dry °C	Wet °C	Degree Days below 18°C	15 Min. Rain mm	One Day Rain mm	Ann. Tot. Ppn. mm	Ground Snow Load kPa Ss	Sr	Hourly Wind Pressures 1/10 kPa	1/30 kPa	1/100 kPa	Seismic Data Za	Zv	Zonal Veloc. Rat.v.
60	Whitehorse	655	-41	-43	25	15	6900	8	40	275	1.7	0.1	0.28	0.34	0.42	2	4	0.20
68	Inuvik	45	-46	-48	25	16	10050	5	55	425	2.1	0.1	0.39	0.55	0.76	1	2	0.10
61	Fort Simpson	120	-45	-47	27	18	8000	8	70	360	2.1	0.1	0.30	0.37	0.46	0	1	0.05
62	Yellowknife	160	-43	-45	25	17	8500	5	55	275	2.0	0.1	0.34	0.43	0.53	0	1	0.05
69	Cambridge Bay	15	-45	-46	16	13	12000	3	35	140	1.5	0.1	0.41	0.50	0.60	0	0	0.00
64	Baker Lake	5	-45	-46	21	15	11000	3	50	260	2.7	0.2	0.42	0.50	0.59	0	0	0.00
74	Resolute	25	-44	-45	11	9	12600	3	25	140	1.6	0.1	0.52	0.63	0.77	2	1	0.05
63	Iqaluit	45	-40	-42	16	11	10050	5	55	425	2.7	0.2	0.39	0.55	0.76	1	0	0.05
49	Vancouver	15	-7	-9	26	19	2925	10	105	1400	1.6	0.2	0.36	0.44	0.53	4	4	0.20
53	Edmonton	645	-32	-34	28	19	5400	23	90	460	1.6	0.1	0.32	0.40	0.51	0	1	0.05
50	Regina	575	-34	-36	31	21	5750	28	95	365	1.3	0.1	0.34	0.39	0.46	0	0	0.00
50	Winnipeg	235	-33	-35	30	23	5900	28	90	500	1.7	0.2	0.35	0.42	0.49	0	0	0.00
46	Sudbury	275	-28	-30	29	21	5400	25	90	875	2.3	0.4	0.29	0.40	0.55	1	1	0.05
43	Toronto	105	-18	-20	31	23	3650	25	90	820	0.8	0.4	0.39	0.48	0.58	1	0	0.05
45	Montreal	25	-23	-26	30	23	4550	23	85	940	2.2	0.4	0.31	0.37	0.44	4	2	0.10
44	Halifax	55	-16	-18	26	20	4100	18	140	1500	1.7	0.5	0.40	0.52	0.67	1	1	0.05

APPENDIX B

BUILDING ELEMENTS

FOUNDATION TYPES   184
WINDOW TYPES   186
DOOR TYPES   186
ROOF TYPES   187
PLASTICS   189
PREFABRICATION   190

FOUNDATION TYPES

Wood piles: A drill rig puts a hole through the active layer to a depth 4 or 6 metres below the permafrost table. A spruce log is set in an augered hole 300 to 500 millimetres in diameter (butt end down to improve resistance to frost heave) and backfilled with a sand slurry. The slurry below the active layer must freeze back to the pile before any loads are imposed. The pile usually depends both on end-bearing capacity and adfreeze strength between the pile surface and the frozen soil to carry the loads, and on adfreeze strength to resist frost heave loads imposed by the active layer when it refreezes. A coat of grease on the portion of the pile within the active layer minimizes uplift forces due to frost heave. Often untreated, wood piles tend not to rot below ground, where soil temperatures remain low all year. But the portions of the pile within the active layer and exposed above ground may rot given the warmer microclimate in the open crawl space caused by heat loss from the building, and some solar radiation. It is essential that those portions be pretreated against rot. The building's ground floor must be insulated and elevated above ground to minimize heat transfer to the ground.

Wood piles are commonly used in the alluvial soils of the Mackenzie River basin. Being within the tree line these areas offer renewable stocks of suitable timber.

Steel pipe piles: In areas where significant depths of ice-rich permafrost occur, a hole (200 mm diameter or less) is drilled deep into the perennially frozen ground. An open-ended steel pipe of specified wall thickness is dropped into the hole and backfilled with sand slurry. The slurry refreezes quickly. The portion of the pile within the active layer is coated with grease. Both the load-carrying capacity and frost-heave resistance of the pile depend on the adfreeze strength of the bond between the pile surface and the permafrost below the active layer.

In areas where ice-rich permafrost is underlain by a thaw-stable bearing stratum the pile can be placed in a drilled hole, and anchored in the thaw-stable bearing stratum by one of several available means. The space between the pile and the surrounding soil is backfilled with sand slurry. If the pile is designed to be end-bearing, it depends on the anchor, the dead load of the building, and adfreeze bond in permafrost to resist frost heave in the active layer. At points supporting light dead loads, such as floor areas away from exterior walls or columns, and outside stairs and landings, special care must be taken to ensure that the piles are well anchored to resist uplift due to frost heave.

Steel pipe piles are used extensively in Canada in the zone of continuous permafrost beyond the tree line.

Gravel pads: In Canadian communities beyond the tree line gravel pads are used extensively under buildings. The gravel pad supports the building directly by means of a wood post on a wood-bearing pad, or indirectly in combination with a pile foundation, by providing a new, non-frost susceptible active layer above the original ground surface. The emphatic visual and physical presence of the pad makes it a factor in the architectural design of the building—the effect is too often ignored. On large tracts of tundra being developed for housing, a continuous blanket of gravel may be substituted for independent pads.

It is important to minimize disturbance to the ground surface when the gravel pads are placed so that thawing and settlement of the gravel pad is kept to a minimum. Usually the gravel is placed during the summer months, when the material is thawed, ready for compaction. The gravel is placed over the thawed active layer by end dumping, so that heavy equipment need not operate on the native surface. Degradation of the permafrost is minimized.

In principle, the gravel pad must be thick enough to ensure that the depth of annual thaw remains within the thaw-stable pad. If the depth of thaw extends into ice-rich permafrost below the pad, then the pad and the foundations it supports will settle. The thickness of gravel required to prevent thaw settlement typically ranges from 2 to more than 3 metres. Therefore, it is common practice to place extruded polystyrene insulation within the pad in order to reduce the maximum depth of annual thaw. When insulation is incorporated, the thickness of the gravel pad can be reduced to 1 or 1.5 metres. The pad should extend at least 1.5 metres beyond the footprint of the building it supports, preferably more, to cope with the erosion caused by roof drainage. Ground surface drainage must be diverted around the pad.

Ideally the gravel pad is left to settle for a full year before building loads are imposed; in that time the position of the permafrost table rises just above the original ground surface, putting the shallow active layer completely inside the non-frost susceptible gravel of the pad. Some differential settlement can be expected, so foundations bearing on the gravel pad must be adjustable in height to allow annual releveling as necessary. In the absence of routine level checks and adjustments the building may rack and the envelope rupture.

Thick footings made of wood treated against rot are placed just in or on top of the gravel pad, fixed in place by steel pins driven into the gravel. A horizontal wood spreader is placed on the full width of the wood pad, followed by a pair of birch wedges that provide the releveling capability while supporting the floor beams of the building. Metal straps that permit some vertical movement for levelling anchor the building to the wood pad. "Deadmen" may be used to anchor the building to the ground. Made of treated wood and buried below the gravel pad, they are connected to the building by cables.

The ground floor is raised above the top of the pad about one metre and insulated. The resulting crawl space leaves room for leveling operations, and allows the prevailing wind to remove heat radiated toward the ground by the building. The perimeter of the crawl space can be skirted with 100 mm grid steel mesh to prevent accumulation of dead storage items and wind-blown rubbish under the building. (Ice fog from clothes-dryer vents and fine blowing snow may clog finer mesh sizes, eventually stopping air flow through the crawl space.) In areas prone to snow-drifting the crawl space must be designed to minimize obstructions to airflow. The long axis of exposed floor beams and supporting wedges, for instance, should be aligned parallel with the prevailing winter wind. Solid skirting of the crawl space defeats the requirement to ensure the free flow of cold air. Floor beams are cross braced for resistance to lateral loads.

Where footings for an elevated building are buried in a deep gravel pad, a layer of rigid insulation placed high inside the pad thickness, above the footing level, raises the permafrost table to a new position higher than the base of the footing. This arrangement ensures that the ground below the footing remains frozen throughout the year.

Wood pads placed over gravel pads support most older public housing units in northern Canada. Newer public housing is set on piles to eliminate the often forgotten annual leveling operation. Since piling is expensive, and often awkward in terms of construction scheduling, housing authorities have experimented with a new foundation method. It consists of placing an exposed space frame made of metal tubes joined in a three-dimensional grid in the full length and width of the crawl space below the building. Wood pads support the frame at regular intervals, just above the gravel pad. Designers expect the rigidity of the space frame to resist the racking effects of differential settlement. But in a substructure with so many connections, the cumulative effect of joint tolerances results in deflections of two or three centimetres over the length of a typical house floor plan. Deflections of that magnitude may be acceptable for some types of construction. The space frame's busy network of tubes may slow the wind enough to produce some snow-drifting below the building.

Ducted gravel pads: For structures such as parking garages or firehalls, where at-grade access for heavy equipment is required, it is inconvenient and expensive to elevate a building above ground. In such cases the designer can opt for a gravel pad foundation that incorporates a layer of insulation and a horizontal pattern of large-diameter pipes. The pipes have both ends open to the outside. Heat radiated from the building ends up in the pipes rather than in the frozen ground. In summer, closing the pipe ends prevents infiltration of warm air. In some cases it is necessary to use an air blower to achieve the required airflow in the ventilation pipes.

Ducted gravel pads are used primarily for fuel storage tanks, warehouses, parking garages, firehalls, and aircraft hangars.

Refrigeration: The thermal regime in permafrost can be kept stable by refrigeration; the expense of installing and maintaining such a system prevents its use in any but special situations.

"Thermopiles" are more common, particularly in the Yukon. In Alaska they are used to support oil pipelines above perennially frozen ground. They are especially useful where the natural permafrost is extensive but too warm to support a long-term structure reliably. Thermopiles are closed hollow piles filled with a liquid that responds to the difference in temperature between the subsurface permafrost (say -5°C) and the air temperature above ground in winter (say -30°C). The liquid is warmest in the ground at the bottom of the pile. Being more buoyant than the liquid at the top of the pile, the warm liquid rises while the liquid at the top, having been chilled by the outside air, descends. The convection cycle removes heat from the ground and releases it into the atmosphere via radiation fins located above ground at the top of the pile. In summer the convection cycle stagnates as soon as the air temperature around the top of the pile rises above ground temperature: the warm liquid inside the pile does not cool again until the fall.

A "single-phase" thermopile contains a liquid that remains a liquid (no change of phase) and functions simply in a convection cycle. A "two-phase" thermopile contains a liquid that changes phase to a vapour and vice versa within the range of temperature differentials typically encountered. Cool liquid descending to the bottom of the pile boils on contact with the heat from the relatively warm permafrost; the resulting vapour in turn recondenses at the cold top end of the pile, releasing heat into the winter air.

Buried footings: Like piles anchored in permafrost, ordinary spread footings can be placed below the permafrost table where they freeze into the soil. The problems are the same: preventing building heat from reaching the ground by conduction through the footing material or by radiation from the ground floor. Unfortunately, the excavation necessary to place the footing will be wide, exposing the permafrost to the elements for extended periods of time followed by another wait as the ground refreezes around the footing. The waiting can be prohibitive at latitudes with short construction seasons.

WINDOW TYPES

Awning window: Hinged at the top, opened outward at the bottom by means of levers and/or handles linked to more levers by worm screws, these have a tendency to freeze, open or shut. The operating hardware is overstressed (or underdesigned), resulting in breakages that are almost never repaired; children take special pleasure in operating such windows repeatedly, with predictable results. In most designs the operating hardware extends from the warm side of the window to the cold side via an over-sized hole, resulting in air leakage; the hardware metal forms a thermal bridge to the outside and becomes coated with hoar frost.

Bay window: By definition the window glass in a bay window is placed remote from room air circulation patterns, meaning that inside air tends to stagnate against the glazing, where it loses its heat, causing the glass temperature to slip below the dew point temperature of the room. Even if air circulation is forced into the bay at the cost of extra equipment and noise, the installed mechanism frequently fails to give the broad sweep necessary to keep the entire glass surface adequately warm. If a curtain or blind is added to the bay window, the situation deteriorates further, since warm room air now has zero chance of circulating against the window glass. Condensation ice forms on the glass.

Casement window: Side-hung, these windows have the same problems as awning windows. Additionally, the open window leaf is exposed to buffeting by the wind, resulting in hardware failure and vibration noise.

Sliding sash window: Whether vertically or horizontally sliding, these windows are extremely leaky—they ventilate the room even in the closed position. In situations where room air is overheated and relatively dry, wooden sash windows do not freeze shut: the wood is an insulator and so much hot air is escaping that the window edge remains above freezing. Freezing at window edges can be expected at normal room temperature, particularly if the sash is made of metal. In the case of multiple sliders, air leakage is so severe that outer panes become opaque with frost at the onset of winter.

Fixed sashless window: Because this type of window is fixed, the ailments described above for opening windows do not occur: air movement around the edges of the glass is close to nil, eliminating one entire set of perimeter joints (sash to frame). The glazing must be placed near the plane of the vapour barrier to ensure circulation of room air against the glass. A separate vent port provides ventilation. Mounted on transom hardware, the vent port, in the open position, frees all edges to inspection, to removal of ice if any, and to repair or removal of a fly screen. A brass or plastic porthole (made for sail boats), placed at the inside face of exterior walls and coupled with a plastic tube that penetrates the wall, works well as a ventilation port.

Part-fixed, part-operable window: A combination of a large fixed window and a small operable sash window performs only moderately well. Problems already described persist: the operable sash window is still too big in area, a condition governed by the size and proportion of the large fixed window; through-the-frame hardware is used; location and proximity of the operable sash window cause cold air to be blown onto the fixed window, reducing the latter's temperature to below the dew point; because it is usually placed at the bottom of the window opening for ease of reach (the coldest position possible), the temperature of the operable sash often drops below the dew point temperature of the room air.

DOOR TYPES

Exterior steel doors: Provided with a core of rigid insulation and a thermal break between the inner and outer faces, steel sheet doors perform well in a tempered winter environment. Used singly, a steel door outperforms a wood door. However, a steel door tends to bow inward toward the heated space in winter: the outer face contracts in the cold while the inner face, much warmer, contracts much less. As a result, only the middle part of the door makes satisfactory contact with the weatherstripping. Steel doors cannot be modified easily to suit field conditions.

Exterior solid wood doors: Wood reacts dimensionally less than steel to changes in temperature. But wood reacts more to changes in air moisture content. In winter the moisture content of the air outside remains below the moisture content of the heated air inside. The residual moisture in the wood plus any vapour entering the door assembly from the room side migrates toward the outer face of the door. The outer face of the wood door swells while the inner face shrinks, bowing the door toward the outside. Only the upper and lower sections of the wood door make satisfactory contact with the weatherstripping. Wood doors can be field-modified to some extent.

Exterior steel/wood doors: Greenlanders commonly install insulated steel doors faced on both sides with wood. Such doors, twice the thickness of ordinary doors, combine the strength and stability of steel with the visual character of wood. Wood facing may be customized, freeing the door to appear as individual as its owner wishes.

Exterior plastic-faced wood doors: Manufacturers make plastic doors to resemble painted wood doors, with dubious results. Unless specially treated during manufacture, plastics exposed to daylight suffer deterioration by ultraviolet light. Unless specifically designed with sufficient thickness, or with shapes that stiffen the plane, thin plastic facing will crack at low temperatures when struck hard. Plastic doors cannot be field-modified.

Glazing in exterior doors: In a moderated outdoor environment sealed glass in exterior doors functions well. But glass in an exterior door fully exposed to the outdoor environment will not resist the extra warping stresses caused by ice jams and snow blockage. Designers prefer solid doors for this reason, and may provide sight lines through the entryway by means of transparent sidelights.

Revolving doors: Invented to limit the influx of cold air at the base of tall buildings each time a person enters from the street, revolving doors function poorly at high latitudes. High first costs, frequent breakdowns of bearings induced by severe environmental stress, and the scarcity of trained mechanics to repair them have delayed general acceptance.

Interior doors: Wood doors can be modified or adjusted in the field with ordinary carpentry tools. Being manufactured in a controlled humidity environment, wood doors react dimensionally on contact with the dry air of heated buildings at high latitudes. They develop new internal stresses, and warp, as wood members inside the door lose moisture at different rates.

Overhead rolling doors: Overhead doors, at high latitudes used mostly for garage entrances, leak air and vapour at a great rate. Their geometry—many long slats joined by hinges combined with poor fit between the building structure and the perimeter of the door—ensures leakiness. Powder snow routinely enters at bottom and sides. Not surprisingly, they react poorly to accidental impacts.

ROOF TYPES

Conventional assembly with MBM: "Conventional" distinguishes a roof assembly from an "inverted" system; designers place the waterproofing on the outside, fully exposed to the elements, and place insulation and vapour barrier on the inside, as shown in Figure 1-26. Modified bitumen membrane (MBM) for cold climates consists of an asphalt modified with styrene-butadiene-styrene rubber and reinforced with polyester fibre mesh or glass fibre mat. Mineral granules coat its exterior surface as protection against ultraviolet light. MBM is a strong sheet material, delivered in rolls, that remains flexible at temperatures as low as -40°C. Installers partially melt the surface to be adhered with a propane torch to attach it to a substrate or to another layer of MBM. MBM has been used on sloped roofs in Greenland for a generation, and in southern Canada for about twenty years. Its use in northern Canada began in the early 1980s.

A concealed layer of MBM fully adhered to an uncorrugated deck on the warm side of the insulation doubles as air and vapour barrier.

In Canada, MBM's newness initially worked against general acceptance. Roofers unfamiliar with the material tend to assume a bond has occurred if the contact surfaces show signs of melting just prior to contact. Unfortunately, at low ambient air temperatures, early signs of melting do not guarantee a proper bond between the MBM top sheet and the base sheet in a two-ply roofing application. In the presence of a cold breeze the contact surfaces of one or both sheets may be insufficiently heated for a proper, full-contact weld. Worker fatigue (waving a heavy torch like a wand for ten hours a day saps upper body strength) exacerbates the situation. Judgment as to whether or not the material has bonded properly fails, especially as poor bonding and good may be difficult to distinguish by casual inspection. The poor bonding results in air pockets, ridging of the top sheet, and probable inclusion of water then or later between the sheets, conditions that lead to early failure. Once the roofers understand the problem, poor bonding can be avoided. On roofs with steep slopes and solid decking, a single extra-thick MBM top sheet may be sufficient to waterproof the building.

The overall performance of MBM to date appears satisfactory. However, since existing MBM roof installations in northern Canada are too new to have aged the length of a predicted lifetime—twenty years—the final reckoning still lies ahead.

Inverted assembly with BUR: Invented about forty years ago with the advent of moisture-resistant foam insulation, this assembly has the waterproofing membrane below the insulation rather than above. The built-up roofing (BUR) spends its service life on the warm side of the insulation, rather than fully exposed to the elements. (BUR, strong and flexible at room temperature, quickly loses flexibility as the service temperature drops.) Roofers lay foam insulation boards over the BUR and cover them with a permeable sheet of plastic as protection against ultraviolet light attack and contamination by gravel fines. They then lay clean gravel ballast over the protection sheet. Some precipitation will flow to drains across the protection sheet and some will seep through the sheet, and through the joints between the foam insulation boards, until stopped by the waterproof BUR; here heat loss from below warms the water that flows to drains along slopes in the structural deck. Minimum slope should be 1:25. Even with positive drainage there is a risk that the insulation board will float if water backs up behind a clogged drain outlet. Note that the BUR, being impervious to vapour and placed at the warm side of the insulation, acts also as the vapour barrier (and as the air barrier if fully adhered to a rigid deck).

Common in southern Canada, the inverted roof system is rarely used in northern Canada despite good performance. The system uses familiar materials well and comparatively cheaply. The main drawbacks (sufficient to make this roof system uncommon in small, remote communities) are: unavailability of clean, graded gravel for ballast in most settlements; installation must be completed in good weather (late fall is too late); heat energy lost to precipitation drainage at membrane level; the difficulty of locating and repairing leaks concealed by insulation and ballast; a stronger roof structure to support the extra weight of the ballast—the thicker the insulation the more the ballast. It remains a good solution for large, nearly flat roofs in urban centres.

Substitutes for BUR in an inverted system include: modified bitumen membrane (MBM); EPDM rubber; polyvinylchloride (PVC). Note that PVC roofing membrane, being pervious to vapour, must be accompanied by a vapour barrier.

Inverted roof systems tend to fail first at exposed edge conditions. At the edges, membrane turned up (away from the warmth of the building) into the weather above the insulation and gravel ballast sustains greater expansion/contraction forces due to temperature change than the roof membrane proper. Ensure that membranes at edges are insulated, and shielded from ultraviolet light.

Conventional assembly with sheet metal: The designer can select an all-metal system or a wood structure assembly with sheet metal waterproofing. The all-metal roof consists of a sheet metal liner fixed to the main steel structure as the ceiling, a vapour barrier above it, glass fibre insulation and corrugated or standing seam sheet metal roof pre-finished at the factory. The most common problems with this assembly are: air leaks, low slope, and insufficient allowance for expansion-contraction. Sheet metal, by definition, comes with lap joints. The joints cannot be effectively sealed against air movement. Often the vapour barrier (polyethylene film), expected to do extra duty as an air barrier, deteriorates when continually flexed by wind buffeting the building. The premanufactured edge details leak air too, to the point that fine powdered snow can enter the roof assembly and eventually melt. Designers who are tempted, for non-roofing reasons, to reduce the slope of a metal roof to the absolute minimum and to compensate by increasing the lap distance of the joints, always disappoint their clients. (The really hopeful designer will place a bead of sealant at laps also.) The difficulty remains that the sheets are always in movement: having a high coefficient of expansion, sheet metal reacts to every passage of sun, shade, and air temperature change by expanding or contracting, and so making lap joints that are permanently unstable. Sealant sandwiched in these moving joints will fail under shear stresses during the first year. These less-than-perfect joints let in wind-driven rain, meltwater, and snow. Further, some sheet metal profiles are susceptible to denting under occasional foot traffic. This obviously deteriorates the lap joint situation further. Only by radically increasing the slope and by adding a functioning air barrier will this major defect of metal roofing assemblies be resolved.

Since metal sheets expand instantly when warmed by a shaft of sunlight, they strain noisily against screw anchors or clips (screw holes increase in size in the process). They also shift in a high wind, sucked in and out by the rapid changes in air pressure. The buffeting noise reverberates inside the building and individual sheets will eventually blow off if unattended.

Achieving a permanent waterproof bond between sheet metal and penetration details like vents and skylights poses a problem, as does making water drain past such details when the roof corrugations or standing seams block the flow.

Bad weather affects construction of all-metal roofs less than most other systems, provided that installers prevent snow from entering the insulation. In practice, designers tend to reserve all-metal roofs for garage and maintenance buildings, where interior comfort and convenience do not head the list of performance requirements.

A sloped conventional roof assembly made of wood structural members and glass fibre insulation topped with sheet metal holds more promise, provided that expansion and contraction of the sheets is accommodated by expansion clips.

Conventional assembly with asphalt shingles: With a good slope and a working air barrier, shingles provide a cheap, good roof for housing in areas within the tree line not affected by high winds. Anywhere else the damage caused by wind uplift, both in removing all or part of a shingle and in allowing entry to wind-driven rain is too great. Installing shingles in cold weather leads to early cracking around roofing nail locations and imperfect bonding of tabbing cement.

Uncertain performers: Roofs with poor or unknown performance at high latitudes include conventional assemblies waterproofed by exposed PVC (ultraviolet attack, low-temperature brittleness, poor footing in winter), exposed EPDM rubber (ultraviolet attack, shrinkage); asphalt selvage roll roofing (low strength).

PLASTICS

For forty years chemists have showered the construction industry with a bewildering array of special-purpose plastics. Since the industry cannot fully integrate new materials without first having them tested in laboratories and in field conditions, the evolution of plastics in building has been chaotic. Survival of the fittest material is complicated by spotty market research, competition with cheaper established materials, company diversification and marketing strategy, and formidable start-up costs for assembly lines. The life of a plastic product on the market may be ten times shorter than its service life in a building, thereby complicating future repair and replacement.

Adding severe environmental conditions to this uncertainty explains the truly accidental incursion of plastics in high latitude building design. The failure of certain roofing membranes, insulations, and window frames and the abrupt withdrawal or restriction of others (urea formaldehyde and polyurethane insulations) in recent years now seems to have been inevitable. The impact of weathering, particularly the ultraviolet component of sunlight, alters the molecular structure of the plastic: useless properties replace useful ones. (The loss of special properties may be rapid, but the residue may persist for generations—witness the wind-blown proliferation of disposable diapers and foam insulation off-cuts on the tundra). The record improves slowly as the service life of various products installed years ago in buildings grows longer, but the layman's inability to judge a plastic's durability by its looks continues to breed confusion.

Plastics are designed at the molecular level: atoms of carbon and hydrogen bond to form long molecules that either lie together like cooked spaghetti (thermoplastics—polystyrene, polyvinylchloride, polyolefin) or are highly cross-linked like molecules in crystals (thermosetting plastics—polyurethane, phenol-formaldehyde, polyisocyanurate). The first softens and melts at low furnace temperatures, while the second remains rigid and chars at similar temperatures. The molecular structure is further modified by addition of fillers, stabilizers, fire retardants, reinforcement, or colorants, to suit specific end uses. A roofing plastic that does not contain an ultraviolet light stabilizer will degrade quickly in sunlight to a material too weak and brittle for roofing purposes. A plastic insulation intended for concealed use shrinks and yellows when exposed to weather.

Open-celled foam insulations absorb water and water vapour easily; closed-cell foam insulations absorb much less. Some foam insulations are more susceptible to flame spread and burning than others; all are dimensionally unstable: they expand and contract significantly with large changes in ambient temperature. Clear acrylic skylights and window panes require special edge details to accommodate this movement. Vinyl floor tiles shrink with age.

PREFABRICATION

Prefabrication of building components surfaces from time to time as the ultimate solution to the world's intractable shelter shortage. The impetus for the idea often dies prematurely or, where totalitarian authority holds sway, prefabricated buildings conceived in haste survive to become slums in less than a generation. Like junk yards, defunct prefabrication projects litter the developing regions of the earth. The world shelter shortage is not a technical problem: its roots are political, social, and economic. At high latitudes prefabrication has succeeded only where money is no object, or where a community is a one-company resource town, in other words, where every other social and technical consideration gives way to achieving a high level of productivity in the shortest possible time.

Prefabrication's selling points are rapid assembly at site, with a minimum of labour, economy of scale, and factory-style quality control of manufactured parts. Casual observers confuse these ideas with low cost, as if the lead time, energy, and capital expended to design, produce, and transport interlocking building components could be excluded from the equation. Given the short building season—late summer (when the materials arrive by ship through unfrozen sea lanes) to the start of winter—rapid assembly is essential, but no longer exclusive to prefabrication technique. Refinements in conventional construction methods have made rapid assembly commonplace. Besides, public policy at high latitudes aims to increase local employment, not to decrease it by purchasing products prefabricated outside the region. And economy of scale fails to benefit any region with markets too small and unsteady to support the capitalization of an assembly line.

Typical prefabrication projects put the design of perfectly interlocking parts ahead of design for human comfort and convenience. The building project develops as a three-dimensional puzzle, rather than as a locus of future human activity. There are other impositions: to control mould or jig costs, constraints are placed on the total number of different parts, which in turn restricts the planning of useful space. Panel size is kept small for handling reasons, often with the result that openings in them for windows are disproportionately small, just as window size in aircraft is determined in part by rib spacing of the air frame. Panel design must include convenience of crating and shipping to site—factors that add nothing to human comfort. Purely technical problems abound: building parts glued together in a factory at room temperature may not fare well at locations where ambient temperatures differ by 60 degrees Celsius from one side of the panel to the other. Metallic outer skins contract severely in winter—they stretch flat—and often buckle when they expand again in the heat of direct sunlight. The solid materials that reinforce panel edges against flexing and incidental damage create thermal bridges precisely at joint locations, where cyclical movement and sealant unreliability already diminish envelope integrity.

Subtle negatives proliferate: prefabricated panels are prefinished on the outside—a simple batch-colour error at the factory discovered at the site long after the assembly line has been dismantled cannot be cheaply remedied. The error survives indefinitely. The finish may last for years but in time it must be rehabilitated—what new coating will adhere for long to a prefinished surface subjected to severe environmental conditions? If the building is to be expanded in fifteen years will the original moulds be available to restart panel production, or will the addition have to bear the cost of a second assembly line? Panels are erected for the first time at the site, but how does a construction superintendent rectify a design or factory error just discovered in the panels without ruining panel integrity? Incidental damage to such panels in transport and storage cannot be repaired at the site to factory standards. In communities that do not possess equipment capable of hoisting panels into position the incidence of damage is greater still.

INDEX

A

absolute zero, 28, 45
absolute humidity, 145
acoustics, 29, 70, 106
active layer, 27, 34-35, 114-119, 121, **155**, 184
adfreeze bonding, 118-119, **155**
air barrier, 48-49, 52, 81-83, 108, 122-125, 128, 133, 135-136, 138, 142-143, 152
air cargo, 77-78
air lock, 70, 109-112
air pressure equalization, 84
air-vapour mixture, 54
Alaska, 3-5, 8-10, 31, 137
albedo, 28, **155**
Alexandra Fiord, 88
Algonkian, 8, 9
Antarctic, 29, 43
Antarctica, 3, 148
Arawaks, 11
Arctic, 5, 8, 12-15, 23, 29, 33, 37, 49, 66-68, 70, 77-79, 88-89, 141, 146, 149, 154
arctic archipelago, 5, 34, 45, 89
Arctic Circle, 43-44, **155**
arctic oasis, 33, 46
Arctic Ocean, 8
Athapaskan, 8-10, 12
Atmospheric Environment Service, 51, 164
atmospheric perspective, 29
attic, 53, 135-136, 138
ATV, 18, **155**
aurora australis, 30, **155**
aurora borealis, 30, **155**
Axel Heiberg Island, 22, 35

B

backer rod, 36, 103
back-to-the-land, 16
Baffin Island, 3, 5, 26
Baker Lake, 3, 45-46, 49, 51
balloon framing, 123-124, **155**
band lifestyle, 14
bare poles, 36
barometric pressure, 54
battered piles, 118
Beaufort Sea, 12
Belcher Islands, 27
Beothuks, 11
Bering Strait, 3
bison, 37
blowing snow, 44, 49, 52, 69, 78, 90, 91, 95, 135, 136, 178
boreal forest, 32, 34, 50, 92, 94, **155**
British Columbia, 9, 18, 26
building envelope, 45-46, 50-53, 55, 80, 82-83, 101, 111, 113, 129, 148, 151, 153, **155**
building program, 71, 105-106
BUR, 135-136, 139, **155**
Burgess Shale, 26

C

cache, 6, 27, 61, 76, 88, 141
California, 4, 23
Cambridge Bay, 17, 51
campsite, 6, 61, 63, 88-89, 92, 147
caribou, 4-5, 7, 13, 25, 27, 35, 37, 48, 61, 66, 76, 88, 94, 132, 141
cathedral ceiling, 41, 53, 138, 151-152, **155**

chemical action, 49
chemical reaction rate, 46
Chile, 4
chinook winds, 51, 90
Chipewyan, 10
Chukotan, 9
circuit court, 17
circulation space, 105, 109, 133
community planning, 64, 93-94
construction cost, 74-75
continuous permafrost zone, 114
cooling rate nomograph, 50
co-operative societies, 15
Cornwallis Island, 89
Cree, 18
creep, 27, 91, 114-115, 143
cricket, 129, 139, **155**

D

Davis Strait, 12, 30
Dawson, 113
daylighting, 70, 110, 126-127, 129-130
days with fog, 57, 181
deadman, **155**, 185
deflector, 70, 97, 99, 111
degree-days, 47, 81, 172
Dene, 5, 8, 12, 15, 36-37, 88, **155**
Detah, 83
Devon Icecap, 10, 23, 33, 50, 54
dew point, 19, 54-57, 123, 126, 128-131, 134, 137, **155**, 186
differential movement, 107
differential settlement, 102, 113-114, 130, 134, 138, 185

differential staining, 49
discontinuous permafrost zone, 114, 119
Disko Bay, 40
Dogrib, 10
door bell, 110
door swing, 112
Dorset culture, 5
double floor, 108
driving rain, 49, 52
drumlin, 27, **155**
drunken forest, 26-27, 118-119, **155**

E
earthquake, 26, 101, 118
Easter Island, 12
education system, 16
efflorescence, 144
Ellesmere Island, 11-12, 23, 26, 30, 88
envelope continuity, 101
erratic, 27, **155**
esker, 27, **155**
Eskimo-Aleut, 9
Euro-Canadian, 8, 13, 16, 63, 72, 146
European, 5, 8, 11-13, 15, 41, 63-64, 66-67
exfoliation, 26

F
fast ice, 24, **155**
Fire Marshal, territorial, 74
forest fire, 31-32, 35
freeze-thaw cycle, 24-27, 34, 45-46, 49, 84, 124, 135, 141, 144
Frobisher, Martin, 5
frost boil, 26-27
frost heave, 114-115, 117, 119, 120-121, 137, 150, **155**, 184

frost susceptible soil, 114, 145, **156**

G
geomagnetic north pole, 30
geostationary satellite, 15, 79
glacial till, 7, 27
glacier, 23-24, 26-27, 33, 141
glare, 41, 72, 127, 129-130, 152
Greenland, 5, 8-9, 11-12, 15, 19, 30, 37, 41, 45, 47, 61, 64, 112, 124, 141, 151-152, 187
Greenlandic, 9
Greenland Icecap, 47
Gulf Stream, 28
Gwich'in, 10

H
heat storage capacity, 56, 143, 148, 150
Home Rule, Greenland, 15
hominid, 2
Hudson Bay, 12, 27
Hudson Bay Company, 62
humidification, 53
hypothermia, 47, 53

I
Ice Age, 3-5, 45, 52
ice lens, 26, 114-115
icecap, 3-5, 23, 27, 30, 33, 45, 47, 52, **156**
Iceland, 5, 31
ice-rich permafrost, 114-115, 117, 119, **156**, 184
ice-wedge polygon, 26, 118
Igarka, 113
infra-red radiation, 42, 46, 128, 136
insulating glass, 128-129
Inuit, 5-6, 8-12, 15, 27, 37, 66-67, **156**
Inuit Circumpolar Conference, 9

inukshuk, 6, 27, **156**
Inuktitut, 9-10, 17, 36, **156**
inverse square law, 103-104
inverted cavity, 66, 70
Iqaluit, 18, 30, 48, 51, 136, 154
isostatic rebound, 3, 5, 26, **156**

J
justice system, 16

K
Keewatin, 13
kitchen, 17-18, 148, 150

L
Labrador, 8, 12, 26
land bridge, 3-4
land claims, 15
lap joints, 102-103, 128, 138, 189
Laurentian Shield, 3, 26
Leaky, Mary, 2-3
lightning, 31
Little Big Horn River, 61
log cabin, 19, 88
log school, 73
London, 12
looming, 30, 33, **156**
low-emissivity coating, 128
Louvre, 154

M
Mackenzie Inuit, 12
Mackenzie River, 12, 25-26, 118, 141, 184
magnetic north pole, 30
maintenance cost, 74-75, 134, 143
Manhattan, 61

matchbox house, 48-49, **156**
MBM, 138-139, **156**, 187-188
mean wind speed, 51
Métis, 8, 10
midnight sun, 77
missionaries, 13, 141
missions, 17, 63, 141
moisture content, 53-57, 80-83, 102, 143-144, **156** 187
molecular diffusion, 56, 82-84, 123, 138
Montagnais, Innu, 8
Montreal, 12, 31, 35, 77
moose, 18, 37
moraine, 23-24, 27, **156**
mud circle, 26
muskeg, 118, **156**
muskox, 33, 47-48
muskrat, 19, 37, 88

N
National Building Code, 46, 50, 53, 74, 182
Navajo, 8
Newfoundland, 11-12
noise pollution, 29
Norse, 5, 12
North Pole, 4, 8-9, 30, 47
North Slavey, 10
Northwest Territories, 7-10, 74, 88, 154
nunatak, 23, **156**
Nunavut, 154
nursing stations, 14, 20, 63-64
Nuuk, 19

O
Old Crow, 4, 44, 78
operating cost, 74-75, 143

orientation, 51, 65, 72, 92, 94,-95, 112, 127
Ottawa, 8, 13
ozone layer, 29, 43

P
pack ice, 19, 24, 77-78, 90
Paris, 154
patterned ground, 26, **156**
pattern staining, 145-146
PCB, 15, 29, **156**
Pei, Ieoh Ming, 154
perimeter-to-area ratio, 99-100
permafrost, 25-27, 35, 91, 113-115, 118-120, **156**, 184, 186
permafrost table, 26, 34, 115, 118-119, **156**, 184-186
permeance, 54
pingo, 26-27, 61, 118, **156**
platform framing, 124, **156**
Pleistocene epoch, 3
polar desert, 33-34, 50, 56
pollution, 27, 29, 35,37
polynya, 24, 88, **156**
portal frame, 107, 124
porthole, 127, 130, 149, 186
positive drainage, 137-139
post and beam, 107, 124
prefabrication, 45, 89, 124, 143, 190
programming, 71, 75, 101, **156**
public housing, 64, 74, 87, 93, 185
pyramid, 154

Q
Quebec, 8, 77

R
rain barrier, 80-84, 123-124
rain screen principle, 84, 123-124
rainfall intensity, 50
raised beach, 5-6, 23, 33, 91, 93
Rankin Inlet, 19, 48, 89, 126
raven, 36-37, 137
reindeer moss, 33, 35
relative humidity, 19, 50, 53-57, 66, 69, 130, 143-144, **156**, 179
residential schools, 13
Resolute, 47, 51-52, 89-90
roches moutonnées, 27, **156**
Rocky Mountains, 8, 26

S
Saint Lawrence River, 12
salamander, 56
sastrugi, 48, 50, **157**
scale model, 96, 104
sculpture, 1, 17-18
screw pop, 102
sea ice, 2, 24, 28, 30, 37,41, 50, 56, 61, 76, 78
sealant shape, 103
sealift, 57, 77-78, **157**
seasonal round, 61-62, 76, 88, 112
shaman, 12, 18
Siberia, 3-5, 8-9
silicone sealant, 36, 137
Skraeling Island, 88
skylight, 19-20, 76, 106, 126, 128-129, 139, 145, 154, 190
smoke alarm, 151
Snag, 46
snow blindness, 41
snow cover, 24-25, 41-42, 47, 50, 52, 77, 90, 129

snow house, 6, 45, 53, 64, 66-69, 80, 99, 109, 113, 122, 126, 132, 135, 149
snowdrift, 23, 25-26, 34, 48-52, 64-65, 67-68, 78, 91, 93-97, 108, 110, 137-139
snowy owl, 36-37
soapstone, 17-18
sod house, 68, 135, 141
soil temperature, 47
solar altitude, 42, 44, 165
solar radiation, 24, 28-29, 41-42, 46, 90, 114, 184
solar shading, 42
solifluction, 26, 91, 118, **157**
solstice, 28, 46-47
South Slavey, 10
space frame, 118, 121, 185
space module, 66-67
space suit, 45-46, 48, 53, 66-67, 80
spiral grain, 142
stack effect, 55, 82-83, 136-137, 148-149, 153, **157**
static electricity, 57
storage, 36, 45, 61, 63, 70, 77-78, 94, 97, 108, 112
subarctic, 7-8, 12-14, 32, 142, **157**
sublimation, 50, 56, **157**
sundog, 30, **157**
sunlight, low-angle, 19, 41-42, 90, 127, 144
sunshine hours, 44, 168
surface-to-volume ratio, 37, 68, 100-101, 142
Surtsey, 31
Sverdrup Glacier, 23, 27, 33, 47
syllabics, 10

T

taiga, 34-36, 77-78, **157**
tent, 6, 23, 32, 36, 45, 48, 61, 63-64, 76, 80, 88, 92, 109, 122, 132, 135
territorial government, 8, 15, 18, 74, 89

thaw bulb, 117, 119-120
thaw-stable permafrost, 114, 117, 119-120, **157**
thermal barrier, 81, 84, 123, 128
thermal break, 100, 130, 133, 144, **157**
thermal bridge, 82, 107, 123, 125, 130-132, 144-146, **157**
thermal resistance, 45-46, 69, 81-82, 122-123, 125, 127-131, 138-140, 142-143
thermokarst, 26-27, 118, **157**
thermopile, 115, 185, 186
Third World, 14, 190
Thule culture, 5-6, 12, 33, 67, 88, 142
tipi, 36, 61, 88, 135, 149
Toronto, 7
trading post, 12, 62-64
training, 14, 16, 18, 74, 98, 136
transpiration, 56
tree line, 5, 8, 26, 35, 49, 52, 65, 90-91, 95, 111, 114, 122, 139, 142, **157**
Truelove Lowland, 34, 51
tuberculosis, 14
tundra, 5, 8, 13, 33-36, 41, 48, 50, 78, 92, 94, 116, 127, 132, 145, **157**

U

ultraviolet radiation, 29, 43, 102, 134, 136, 138, 143, 145, 187
utilidor, 151, **157**
utility-to-weight ratio, 68, 143-144

V

Vancouver, 47
vapour barrier, 53, 80-84, 108, 122-124, 128, 135, 138-139, 143, 145, 152, 188
vapour pressure, 54, **157**, 180
vehicular traffic, 17, 19, 48, 65, 77, 79, 96-97, 113, 144, 152-153
vent port, 127-128, 131, 186
Virginia Falls, 21, 35

W

water flume trial, 96
water margin, 90
water vapour, 19, 45, 47, 49-50, 53-57, 66, 80-84, 122-123, 130, 132, 136-137, 143, 145, 148
water, phases, 50
weatherstripping, 112, 131-134, 187
Western Cordillera, 3
wet bulb temperature, 54
Whitehorse, 31, 47
whiteout, 52, 126, **157**
Willow Lake, 88
wind chill, 49-51, 66, 151, 175
wind porch, 109, 111
wind rose, 57, 94
wind scour, 95, 111
Winnipeg, 45, 51
winter road, 36, 77-78

Y

Yellowknife, 3, 7-8, 20, 26, 35, 47-48, 50
Yukon, 4, 8, 45-46, 113

Z

zoning, 60, 93-94

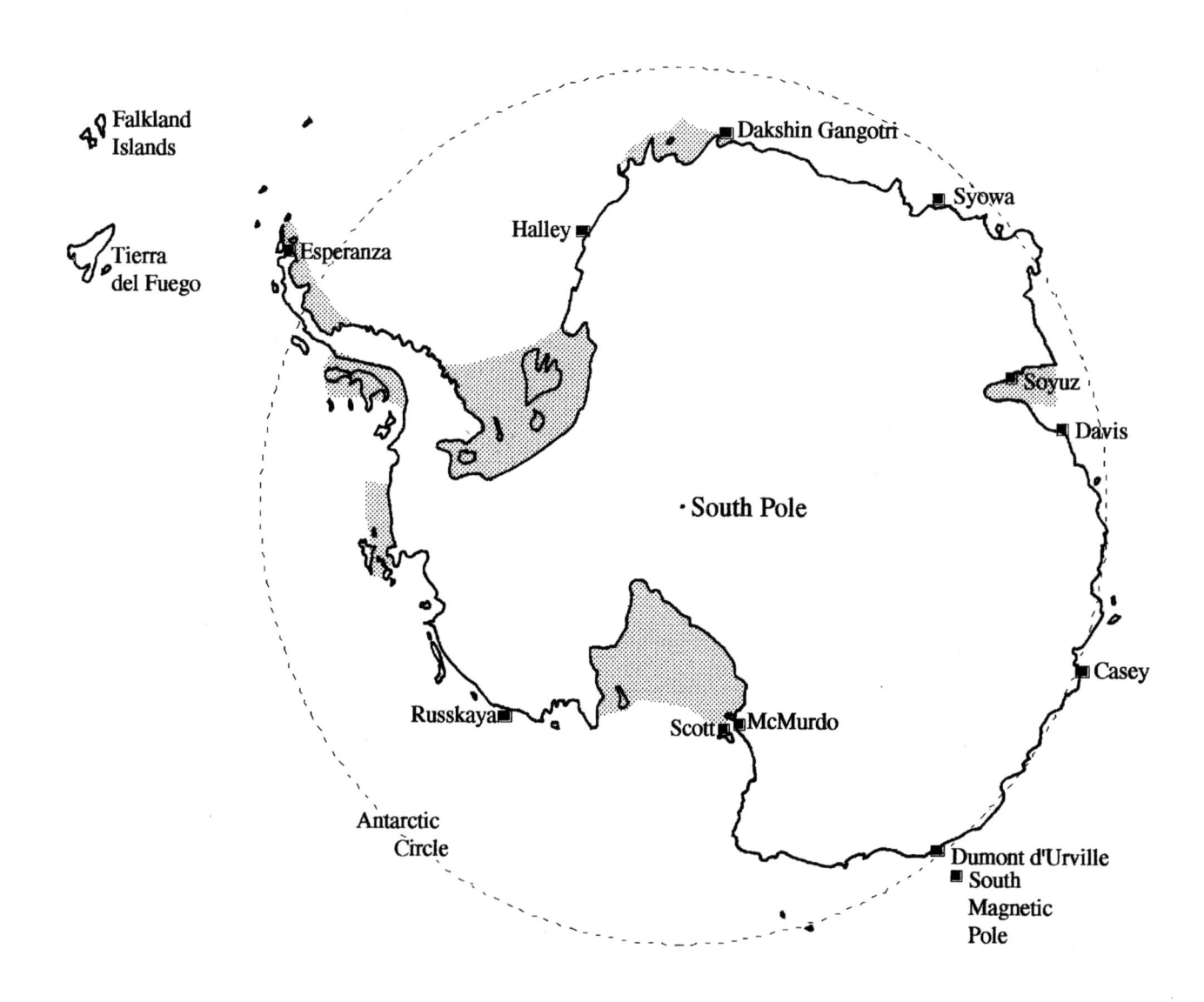